animal days

Published by Bay Books Pty Ltd.
61-69 Anzac Parade, Kensington
NSW 2033 Australia
©1984 Bay Books
Designed by Sackville Design Group Ltd.
National Library of Australia
Card Number & ISBN 0 85835 528 0
All rights reserved.
Printed in Singapore
by Toppan Printing Company

animal days

David Sharp

Edited by
Cathy Kilpatrick B.Sc., Dip.Ed., F.Z.S.

Bay Books

Contents

Introduction	9
The tiger	10
The sidewinder	28
The lemur	40
The giraffe	50
The kangaroo	64
The African hunting dog	78
The gorilla	92
The hippopotamus	102
The hummingbird	117

Introduction

The lives of most wild animals are often mysterious to Man and none is completely understood by even the most knowledgeable expert. However, enough is known to stimulate our curiosity and make us ask such questions as: 'How do animals spend their days?' and 'What would it be like to be a Bengal tiger, a kangaroo or a giraffe?' A closer look at a few of the world's most interesting creatures, described in this book, may answer these questions.

In a world of shrinking animal resources, which is suffering the gradual erosion of its genetic treasures, curiosity about wildlife and concern for its future is more important than ever before. The animal kingdom is confronting great dangers while, paradoxically, receiving more scientific attention and interest from conservation groups than at any other time in its history. By attempting to understand each animal's behaviour, physical attributes and how it adapts to its particular environment, we can help to protect our wildlife and to conserve the most endangered species which are threatened with imminent extinction.

This book focuses on the animals and wildlife of the world's hot regions. Native animal communities have adapted to this environment in wonderful ways; not only does the physiology of different species change to meet the challenges of the habitat but animal behaviour also shows a distinct pattern of adaptation to heat. The most richly varied animal communities in the world are to be found in its hot and warm lands, ranging over such habitats as the tropical rainforests of West Africa, the arid deserts of Central America, the savannahs of southern Africa and the jungles of South-east Asia. An evolutionary balance, maintained by nature for many centuries, is gradually being eroded by Man, and severe damage to one species can affect the populations of other species and the health of the whole animal community. Sadly, two of the beautiful creatures featured in this book, the tiger and the lemur, both of which have successfully adapted to the demands of their environments for thousands of years, are now in danger of extinction.

This book focuses on nine animals, describing a single day in the life of each animal followed by relevant information and scientific data. The day outlined may be unusual or typical as the animal experiences birth, death, flight from a predator or just its routine hunting and grazing in order to survive.

Of course, none of the animals experiences the full range of its activities in one day, but the day described does offer a characteristic sample of its adventures. In one story, a pack of African hunting dogs chase and kill a gnu on their evening hunt, in another the leaders of two gorilla troops challenge each other to a ritualised fight while their fellows watch from a safe distance. By exploring the animal world in this way, we can understand these beautiful animals better and see the world through their own eyes.

The Tiger

The sun moved from its searing zenith and the shadows grew millimetre by millimetre across the quivering ground. Scarlet blossoms blazed on the sunbathed bombax trees and, despite the heat of the afternoon sun, the forest was green with February foliage. The hardwood sal trees whispered in their tall tops as a gentle breeze wafted through their broad leaves, and stripes of shadow flowed across a forest floor which was alive with the humming of insects.

A red-gold body rolled deeper into the dense shadow thrown by a karanda tree. The tiger, a young female, stretched out on her back, showing the soft creamy-white of the fur on her belly. Her paws, seeming a little large for the rest of her body, waved lazily in the air. She sighed and swept one of them past her right ear to rid herself of the flies that buzzed there, feeding on the wax clinging to the snowy hairs that lined the ear.

Evergreen leaves curtained the tiger's lair which faced out across a narrow glade on the sides of a nullah (ravine). The sides of this cleft were a tangle of thorns and vines, but an old deer track led through them down to the stream and pools at the bottom. Little of the water could be seen beneath the rich carpet of water lilies and weed, which occasionally swayed as a hidden movement flexed the green, pink and white surface.

The young tigress left the thorny shelter of the koronda tree, extended every muscle and tendon in a long, long stretch, and swayed slowly down the track towards the water. Near its margin she paused, raised her head and blew gently through her mouth and nostrils. She scanned the reeds nearby and looked further out to the lily pond. Her head drooped and she moaned low in her throat, a strange crooning note. As she watched, a path slowly opened behind a golden, black-striped head which emerged and idly shook itself free of the clinging weeds. A fully grown tigress left the cool water where she had dozed for the last half hour. She scrambled easily up the steep bank to lie on a grassy ledge above the water. There, she began to lick herself clean. Her offspring, now two and a half years old, wandered across to her and dropped heavily down beside her.

The young female had survived her sibling, who had died slowly and painfully a year ago. He had been learning to hunt with his mother when he had seen a movement to the side of the jungle trail that they were patrolling. As he bounded into the thicket he came upon a porcupine, its quills rattling angrily. He pounced, driving his milk teeth into the porcupine's back and pinning it down with a forepaw. He uttered a furious and agonising howl as the porcupine drove its quills into his face and sensitive pads. Backing clear of the dying porcupine, he brushed his face wildly to rid himself of the lethal spines. His mother, bounding after him into the thicket, with a swipe of her powerful forearm flung the porcupine against the trunk of a pipal tree. The prickly animal fell to the ground on its back, quite dead.

The tigress did her best to draw the spines from her cub's face and paws, but two spines had bent round inside one of his pads, laming him, and another had penetrated deep below his left eye. In his frantic attempts to brush it away by rubbing, he had driven it in deeper still.

As the days passed, the cub's wounds festered, despite the careful licking that the tigress lavished on him. The porcupine spines,

Opposite page:
The magnificent tiger, with its graceful, sinuous body, powerful muscles and speed of movement, is ideally equipped for its role as a carnivore. Slimmer and more slenderly built than a lion, it is the largest of the big cats. Although its distinctive striped golden fur makes it instantly recognisable, it is, in fact, an excellent camouflage when moving in the sun dappled jungles and forests that are its natural habitat. However, six of the eight races of tiger are now in danger of extinction owing to indiscriminate hunting of this beautiful animal and the destruction by man of the forests in which it lives. A solitary animal, the tiger often stalks its prey over many kilometres before it rushes at it and springs onto its back, breaking its neck and killing it with razor-sharp teeth.

slowly worked their way deeper into the cub's flesh at a rate of about 25mm a day. He began to lag behind in the hunting expeditions and his appetite diminished, but the tigress continued to care for him.

One night she left the lair to eat the carcass of a pig which she had killed the previous afternoon. She had hidden the pig in a thicket further than usual from her lair. The female cub accompanied her but the young male tiger made no move to follow them. It was not until the females were eating the pig that the high-pitched yelping of a pack of wild dogs had alerted the tigress to her cub's danger. She bounded through the dark forest but arrived near the nullah to find the small pack had killed the cub and already eaten much of him. Leaping in to scatter them, she had brained one dog with a clout of a forepaw, and her great canine teeth crunched through the spine of another. The remaining dogs had broken from the kill and fled. The pack, no more than six strong, was a small one and their bellies were full or they might have stayed at a safe distance to stalk and encircle the tigress. The female cub ran up a minute later, sniffed the nearest dead dog and circled the area of the conflict while the tigress nuzzled the body of her dead cub. Finally, she had raised her head and the forest froze into a dreadful silence as she roared her fury, threatening any living thing that should dare come near her that night. Even her other cub slunk, whimpering quietly, back to the lair.

Now, a year later, the two tigresses spent more and more time apart. The cub was maturing fast. A few more months and she would be ready to mate and bear her first offspring. For more than a month now her mother had been on heat, ready to mate. She roared her frustration at the absence of a suitable male every night, but no tiger had been within the two kilometres range of her roar, nor had a male smelled the scent signals she had laid in her territory.

Opposite page:
This tigress is carrying a young cub in her mouth as a domestic cat would transport a kitten. Male and female tigers come togther to mate only during the two weeks of the year when the tigress is in season. Fights to the death over possession of a tigress are not uncommon among males during this time. After a gestation period of about 105 to 115 days, a litter of three to six cubs is born. They stay with the tigress for two years, learning the skills of hunting and survival in the jungle, and therefore female tigers tend to mate and breed only once every three years.

This young tiger cub is a white tiger of Rewa and may be born in the same litter as normally marked cubs. Unlike them, its coat is more creamy with less distinctive stripes, its eyes are blue and the nose and pads are pink. Tiger cubs are born blind and helpless, their eyes opening after two weeks. Four weeks later they are weaned and by the age of seven months they have learned to kill prey although they generally stay with their mother until they are two years old. However, the mortality rate among cubs is high and usually no more than one or two cubs survive to adulthood. Sickly cubs are weaned out of the litter by the tigress and some cubs may fall prey to snakes or porcupines.

It was three days since the tigress had eaten a substantial meal. She had licked up a few maggots from a rotten kill left by her cub, and she felt hungry and tense. As she finished cleaning her fur, her tail gave irritable side-to-side twitches. The cub, recognising the danger signals, kept well away and lay quietly on her side, watching. Only the occasional upward flick of her tail betrayed her silent presence in the reeds at the edge of the pond.

The tigress climbed slowly out of the nullah and walked south through the sal and mahua trees. Every two hundred metres or so she paused, raising her tail and sending a jet of urine, mixed with fluid from the scent glands near her anus, arcing two or three metres through the air to mark the undersides of leaves near her route. Once, she turned and sniffed her mark. She turned her head away, folding back her lips in a grimace. Her tread was soft on her rounded, cushioned pads. The only sound was the occasional soft moaning that came from her lips. Her head dropped low and her eyes, yellow in the afternoon light, searched the spaces between the trees for tell-tale signs of prey.

She was near the edge of the trees, where in an earlier season villagers had burned the elephant grass and bushes to make a plain where the domestic buffalo might graze. She paused. There, 200m away, stood a chital doe. She moved a few metres towards the tigress, looking for succulent grass and leaves. Fifty metres beyond her clustered the rest of the deer herd, about nine in number. Her russet coat was dappled with pale spots like the sun-flecked shadows that flickered about her.

The tigress sank slowly down until her white belly brushed the ground. The piled muscle hunched at her shoulders and she re-

During the day, many tigers lie low in the shade of caves, rocks, long grass, trees or rivers. Because of their northern Siberian origins, they do not relish the heat and may even be found in swamps or basking in shallow pools. In this way, they are well camouflaged from other animals, their stripes blending into the shadowy background of the sun-dappled forests and undergrowth. Their markings also serve a useful purpose when stalking prey as they provide the tiger with essential camouflage cover until it is sufficiently close to rush the prey animal and kill it.

This Siberian tiger is specially adapted for its life in the harsh Siberian climate. It differs only in size, length of fur and markings from its Bengal relative. It is heavier in build although the same length from nose to tail with a warmer, more shaggy coat and paler stripes. In summer the coat darkens from the pale yellow of winter to a rich tawny gold. Although only about 200 of these magnificent animals survive in their natural wild surroundings Siberian tigers have been bred very successfully in zoos throughout the world.

mained perfectly still. Gradually she flowed forwards. Nothing was too small to offer her cover; even a tuft of grass was enough to make her seem to vanish. Then the bright red-gold and black of her coat slid soundlessly forwards a few metres to a fold in the ground, where she submerged from view – only from the chital doe's view though. In the trees high above the huntress and her prey, bright monkeys' eyes watched the hunt. All at once the forest rang with the guttural alarm calls of a family of langurs. The monkeys leapt wildly from branch to branch over the tigress's head, making her start up in astonishment. The surprise quickly turned to fury as the chital deer stampeded away from her. She roared savagely at the langurs which swung and leapt away through the treetops as if they inhabited another element, whooping their derision at the earthbound terror roaring furiously below them.

Two hours later the tigress was stealthily scouting the edge of the tree-line. The light was fading into the golden pre-dusk glow of the late Indian spring; and out on the plain, near a tall patch of elephant grass, the forked horns of a sambar stag lifted into view. A few metres beyond him three does grazed, quiet but alert. The sambar was out earlier than usual.

The tigress sank to the ground, utterly still, watching the peaceful scene. She stalked towards a thorny ber bush, slowly, slowly, until she could see the sambar stag clearly. There was no way in which she could close the distance without being seen. Hidden behind the bush, she gradually raised her head and made a curious sound: 'dhank, dhank'. The stag was alert at once, his head pointing towards the bush. The tigress repeated the sound and the stag trotted a few paces forwards. He was listening to the alarm call of the sambar –

or at least, that is what the sound resembled – and his curiosity was aroused. He was a fine beast of about 240kg and stood 150cm high at the shoulder. His horns swept upwards and back 90cm from his crown, and two pointed tines sprang forwards from their base. The early evening breeze tippled through the long hair that fringed his greyish-brown coat at the neck. He waited, uncertain. There was something odd about the call, but instinct drove him to discover the danger so that he might know how to react. Behind him, the does stood watchfully.

'Dhank, dhank'. The sound of the belling sambar drew him forwards at a trot. He was less than a hundred metres away now – not near enough, the tigress knew. She could charge her prey at over 60km per hour, but the sambar would react like lightning and she would not be able to close the gap fast enough before her speed began to die away. The sambar could run fast for many kilometres, but the tigress would run herself out in less than 300m.

She tried her imitation call again. The stag was distinctly uneasy now. Perhaps he found the sound thinner in quality than the rounded belling of his own kind, but he still wanted to discover where the trouble lay. He trotted forwards again. Fifty metres, forty – the range was nearly right. The tigress sank lower to the earth, her long hindlegs gathered underneath her, chin touching the ground, lips puckering back to leave the curving canine teeth gleaming clear. At last the scent of the stag drifted to the tigress's nostrils. Low to the ground, windborne scent was slow to catch her notice.

This tiger, patrolling a shallow river bed in search of food moves almost silently with an undulating stride. It walks in a digitigrade manner with the toes and balls of the feet only touching the ground while the palms and soles are raised clear. Tigers often wait at waterholes where prey animals gather to drink and pounce on their prey when they appear. However, despite its reputation for being a powerful and effective hunter, the tiger often goes for several days without food in the wild and must be clever and stealthy in its hunting. Tigers usually patrol their individual territories, marking the boundaries with scratchmarks.

This tiger is dragging its kill to the nearest watering spot to be eaten at leisure. Tigers usually drink large quantities of water when feeding and therefore they may pull their kills for several hundred metres to a pool or stream before eating them. They have been known to lug heavy cattle and buffalo that would take a dozen men to lift over considerable distances in search of water. If the whole carcass is not eaten at a single sitting it may be concealed behind boulders or in undergrowth until the next meal. Although it is essentially solitary in its habits, the tiger sometimes shares a carcass with its mate or other tigers, the group taking it in turns to feed and cleaning themselves afterwards.

It was enough. All the tension of the long wait broke. The ber bush seemed to explode as the tigress's spring drove her body through the air, smashing its way past the twigs to one side of the bush. She did not touch the ground again until she had flown five metres towards the stag with an ear-splitting roar. The stag seemed to rise off the ground and hang in the air for a terrible moment as he saw the blazing ferocity that streamed towards him. He drove his hooves into the ground with desperate force, slewing his body away from the carnivore's rush. The dust of the soft earth flew and the tigress was through it before it had time to settle. Her concentration was appalling. Her eyes seemed to draw her home on her prey until she was close under his left flank.

One last spring shot her into the air. The stag's wrenching swerve away from her golden blur faltered in a patch of soft ground. A great forepaw circled over his straining neck and five sharp claws slapped deep into the right side of his neck. Almost at once, the tigress drove her upper canine teeth into the stag's throat and closed her jaws to force in the lower canines. The great stabbing teeth crushed the stag's windpipe and tore the jugular vein, but still he staggered forwards.

The tigress vaulted over her prey, her teeth remaining closed in his neck. This was too much for the powerful stag. The heavy tigress circling over his back dragged him off his feet and he fell on his right side. With her strong teeth, still buried in his neck, the huntress twisted the head of the deer so that his own weight broke his neck as he hit the ground in a cloud of golden dust.

The tigress did not move, nor did she relinquish her grip until the stag's hind legs kicked in their death throes. The lithe huntress sucked her flanks in and out as she panted. The does had fled towards the middle of the plain where they could see the tigress's approach in ample time to move away. She casually swept the claws of her right forepaw across the stag's haunch to lay open the rich muscle meat under the skin. A few mouthfuls satisfied her for a while and, pausing to rest, she scanned the plain around her.

The lair was about two and a half kilometres distant, and the light was turning the plain to a reddish-gold in the beginning of the sunset. The tigress tugged her prey towards the distant nullah, making little of the stag's great weight. After half a kilometre of this progress she shifted her grip and, seizing its neck in her jaws, she swung the carcass across her shoulders. With her head braced high, she made better time through the more tangled cover of the forest. She struck the nullah about a kilometre south of her lair and dropped her burden in a thicket near the bottom. Carefully, she scraped earth and leaves about the stag's body until it was perfectly hidden from sight. She padded silently away to plunge into the stream at the bottom of the ravine. She ducked her head under the surface and swilled the water to and fro between her jaws. Leaving the stream, she lay up on the north bank in the fading light, just in sight of her cache, and began the pleasant process of licking and cleaning. The female cub slipped through the undergrowth to peer across the stream at her mother. The tigress growled deep in her throat and the young female lay down at a safe distance.

The tigress did not remain settled for long. She began to pace about a small area of open ground making mewling noises. These rose in a long, piercing yowl as she called for a male. For many nights she had called out this furious invitation but there had been no response. Once, when she had been a cub, a female's cry would have drawn the attentions of two or three males at a time, but now traps and hunting had taken too great a toll and there were few mature tigers still in her part of the forest.

The night was bright with moonlight, strong enough to throw black shadows across the glade. The tigress trotted purposefully to the hidden carcass. She carefully scraped away the earth and leaves. Suddenly the night seemed to crack open as a shattering roar ripped through the forest. The tigress for a moment was utterly still. She padded away from her kill towards the middle of the clearing. Sending out a roar of her own that inflected upwards in greeting, she lay down in the moonlight. A newly fallen sal leaf crackled nearby. From the edge of the trees, twin points of emerald light swung from side to side as a male tiger stepped out slowly.

The female greeted him with deep rumblings in her throat. She stayed pressed to the ground with her ears laid well back. The musky smell of the tiger filled her nostrils as he walked gracefully towards her, his toes curling inwards and his powerful shoulders rolling. He made a cavernous 'grr-aounch' sound deep in his chest. He rubbed his chin gently over the back of her head and she, rolling onto her flank, gave him a playful bite. He blocked the bite with his own teeth, and for several minutes they fenced with these terrible weapons, clicking them together in the cold moonlight which turned their coats to eddies of grey smoke.

She lay on her belly as the tiger mounted her back. The glade was ringing with snarling roars. The tiger gripped the female's scruff in his teeth, dragging it upwards as he mated with her. A series of furious jerks, and the copulation was over in less than twenty seconds. The female twisted round with a vicious snarl and the male backed off her uneasily.

While the tigress licked herself clean and tidy, the male circled her, gently rumbling in his throat. He was a well-grown animal, his coat marred by a jagged scar on the left flank where a bull gaur, a formidable buffalo-like animal, had caught him with its horns.

The tigress stretched carefully and trotted towards her hidden

Opposite page:
This tiger is drinking from a forest stream. Because they dislike intense heat, tigers often sit in shallow pools and swamps as well as caves and long grass to keep cool during the hottest part of the day. They are strong swimmers and in times of flood they may take to the water, swimming powerfully from island to island in search of food. Sometimes they feed on fish, turtles and crocodiles as well as their usual land diet of deer, monkeys, wild pigs and antelope.

The Siberian tiger is lighter in colour than the Bengal tiger and has fewer stripes. Fossil evidence has demonstrated that the tiger originated in Siberia and spread southwards across Asia in the Ice Ages. Thus the Siberian tiger is closest to the original species. With its thick, long, shaggy coat, it can withstand extreme cold and adapt to the snows and bitter winters of Siberia and Manchuria. Although it was once widespread, there are now probably only about 200 of these animals living in the wild. The largest of all the tigers, the Siberian tiger may weigh up to 290kg and measure four metres in length.

kill. She made a rasping, purring sound in her throat as she approached it. She blew the dirt and leaves from the stag's shoulde before seizing it and giving the whole carcass a thorough shake. A few beetles, many ants and some sleepy hornets fell from the dea animal. The tigress had suffered in the past from the attentions o hornets when they had stung her on the lips – sleepy or not. She la down carefully to eat from the haunch.

She watched the male tiger follow her, pausing every few metre to listen, the white tips of his ears clearly visible in the light of th moon. He sniffed the air intently before slowly walking forwards few more paces, his hind paws usually stepping neatly into the mark made by his forepaws. Thirty metres from the feeding tigress h halted, listened, sniffed the still, cold air and lay gently down t wait. The tigress had opened the stag's flank, eaten the liver an nibbled away some of the meat from the ribs. Since making the kil she had eaten about 20 kilos of meat and she was visibly mor rounded than she had been before the hunt.

She padded towards the stream that threaded its way to the pond Her drinking there did not disturb a wild boar who was alread slaking his own thirst. He could sense that he had little to fear fro so satisfied a tigress. The male tiger approached the kill, grindin his teeth in anticipation of the meal. He too worked from the hind quarters towards the head. He tore out large chunks of meat fror the carcass, shearing them with the special slicing teeth of his pre molars and molars into pieces of manageable size before swallowing

The meal over, he covered the remains for another day and fo lowed the tigress to her lair above the nullah. For several minute he scouted the area. He walked off his restlessness, marking th trees with urine to show that this was his territory too, for as lon

These tigers are mating, the handsome male characteristically seizing the female's neck between his powerful teeth. During the mating season when the tigress comes on heat, she roars and moans loudly to attract a mate, wandering around restlessly until she meets a male tiger. They brush against each other before mating, which is accompanied by noisy snarling and roaring by both partners. They may remain together for several weeks afterwards, continuing to mate while the tigress is on heat, and hunting together.

as he decided to stay. He stopped beside a monsoon-bowed jaman tree and reached up its trunk as high as he could stretch. His claws were fully extended and he raked them down the tree, ripping through the bark to leave oozing stripes on the trunk. The ragged, flaking material at the edges of his claws was scraped by the bark, and a layer of the tree's gum stuck to the claws as they slid back into their sheaths. Finally he lay down to groom himself.

As dawn lightened the sky to the east, the tigress walked along the edge of the nullah. Her cub lay deep in a thicket, watching. The tigress made a soft 'bru-brruuu' sound as she paused by the thicket, and the young female approached her mother warily. The tigress nuzzled her and rubbed her neck along the youngster's flank. She turned and walked towards the stag's carcass, followed by the cub, while both were watched with sleepy interest by the male tiger. The tigress left her offspring eating the remains of the stag and wandered slowly back towards the lair. On the way, she defecated, leaving the faeces exposed but scratching the earth beside them – an indication to any patrolling tiger that the territory was occupied.

The tigers lazed in the morning sunshine, warming themselves luxuriously. The male would probably stay with the two females until the cubs were born, but the young female would soon leave to find territory of her own – already she spent much of her time away from her mother. By mid-morning, the tigers were dozing between visits to the cooling water of the stream. The burly male especially enjoyed making a short leap from the bank into the water to raise a great splash, but as the day grew hotter and midday approached even this activity ceased and they sought the restfulness of cool shade. There, in the calm of the nullah, they would relax until their 'day' began again in the late afternoon.

Distribution

The southwards thrust of the last Ice Age forced tigers, like many other animals, to seek new habitats in warmer climates. Although some animals failed to adapt to new conditions and also the competitors they encountered there, the tiger spread from the Caspian coast to the taiga (coniferous forests south of the Siberian tundra) spreading west from the Sea of Okhotsk and down to the tropical jungles of south-east Asia as far as the island of Sumatra.

Sadly, the encroachment of men, their towns and agriculture (and also their pleasure in killing for its own sake) has slowly destroyed the tiger's habitat (and with it the tigers) over much of Asia. For ideal hunting territory, a tiger needs three things: game, water and cover. With abundant game or domestic stock to eat, a constant supply of water to drink or to cool off in, and dense thickets to provide cover for rest or ambush, the tiger can lead the life for which Nature has so beautifully made him. However, this sort of habitat is now increasingly hard to find.

The tiger's southwards migration, made over many thousands of years, has produced several sub-species of which there are eight that are widely accepted as being distinct. Naturalists hold conflicting opinions over some of the southern varieties, believing them to be variants of one or other of the main types.

Of all the world's tigers, the only race that is not so severely threatened that it appears in the Survival Service Commission's *Red Data Book* on endangered animals is the Bengal tiger; and even that superb beast, despite strenuous efforts made to protect it in some Indian states, may soon be listed as a creature on the verge of extinction.

Species of tiger

The **Siberian tiger** (*Panthera tigris altaica*) is the largest of all the tigers. Once widespread, it is now confined to a few areas of Manchuria, northern Korea and some nature reserves in the USSR. The Siberian tiger's principal headquarters is now the Sudzukhe and the Sikhote Alin nature reserves east of the Ussuri River in the Soviet Union. There are probably only about 200 of these nomadic tigers left.

The Siberian tiger has long outer hair, up to 14cm long, growing through an inner layer of softer, warmer fur. Under the skin, along the flanks and belly, it may have a layer of fat up to 5cm thick. This helps the tiger to resist the bitter cold of the Siberian winters. While the snow is on the ground the tiger's coat becomes yellowish, but when the spring arrives, it darkens to a tawny gold. An adult male may measure over four metres long from nose to tail, and the animal may weigh 270kg. It is the Siberian tiger's greater weight that makes it the largest of tigers; there is little difference in the measurements of length between this sub-species and the Caspian or Bengal tigers. Siberian tigers mate between December and February.

The **Caspian tiger** (*P.t. virgata*) was once one of the most widespread of all the races of tiger. Now with the destruction of the Caspian forests and with man's alteration of the habitat and his over-hunting of the prey on which tigers depend, the Caspian tiger is reduced to a possible few in Afghanistan and perhaps 80-100 in northern Iran.

The Caspian tiger's winter hair is fairly long. Its belly develops a fringe of white hair and it sports a short mane. The hair on its cheeks grows into a marked 'mutton-chop' beard. The tiger's stripes are browner and narrower than those of the Bengal tiger.

Corbett's tiger (*P.t. corbetti*) is lighter than the Chinese tiger, but darker than the Bengal tiger. J Corbett identified this tiger, and it was accepted as a separate sub-species as recently as 1968. It has many stripes and they are set close together.

Bengal tiger (*P.t. tigris*). See *At a glance* panel on page 40.

The **Chinese tiger** (*P.t. amoyensis*), once found in great numbers between the 38th and 40th parallels of latitude, is now reduced to a very small population south-west of Yangtze-Kiang.

The difference between the Chinese and Bengal tigers has never been satisfactorily defined, but the Chinese species has darker colouring. The dark area along the spine tends to continue down the tail and the Chinese tiger is a little smaller than the Bengal tiger.

The **Sumatran tiger** (*P.t. sumatrae*) struggles for survival in the mountains of the northern part of the island. Its numbers are uncertain but hunters and trappers have seriously reduced them since the end of World War II. It is smaller than the Bengal tiger, standing only about 74cm at the shoulder, and it has closer set stripes. The white area on the under-belly is small and the background colour is dark. It has a short mane on its neck and a well-developed ruff of cheek hair.

The **Javan tiger** (*P.t. sondaica*) is one of several sub-species that depend greatly on zoos for their maintenance. In the wild it has been reduced to a desperately small population of little more than a dozen animals. It is a small tiger, its stripes being narrower than those of the Bengal tiger.

The **Bali tiger** (*P.t. balica*) is the smallest of all tigers and is very nearly extinct. Those few that may remain are still sought by hunters in the western tip of the island.

Opposite page:
Some of the different races of tiger are shown here with their varying appearance, size, weight, colour and markings. Although the first tigers originated in Siberia and were built for life in a cold climate, like the present Siberian tiger which has longer, shaggier fur than other species of tiger, the tiger spread southwards across the Asian continent and has adapted to the conditions and climate of its new environment. It now ranges over a wide variety of habitats including swampy jungles.

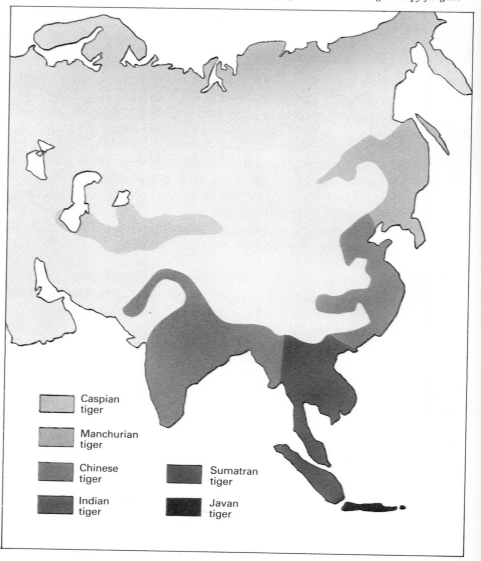

Caspian tiger
Manchurian tiger
Chinese tiger
Indian tiger
Sumatran tiger
Javan tiger

Growth and training

The tigress prepares for the birth of her cubs by choosing a suitable place for her lair. She may already have several places of 'preferred cover' in her territory but, as birth comes nearer, she settles in one place. She piles the floor of the lair with soft leaves and grass to make a comfortable bedding for the cubs when they are born.

After carrying the cubs for 14-16 weeks, the tigress gives birth to two or three cubs. At first they are blind and helpless. After one or two weeks, the cubs open their eyes but remain short-sighted. During this time, the tigress has several problems to contend with: she must hunt over a wider area than usual because animals are often alerted to danger in the vicinity of the lair, vary the times of her hunting expeditions and take extra care over hiding her tracks near the lair so as to conceal the presence of her vulnerable offspring. If she finds the ground around the lair disturbed, she will carry the young by the scruffs of their necks, just like a domestic cat, to a new site.

The cubs drink their mother's milk for about two months before she begins to wean them. When they start to nip and scratch the tigress as they suckle, they are gradually introduced to solid foods. She brings them half-digested pieces of meat and partly decayed meat which is soft enough to be digested easily, but she continues to feed them with her own milk until they are a little over five months old.

At six to eight weeks of age, the cubs begin their education as hunters by following their mother on her hunting patrols. They chase the creatures that stir near their paths, bounding after butterflies, peahens and rodents. They learn to stalk by imitating their mother as she walks silently through the jungle.

Sometimes the tigress makes a kill on her own and carefully hides it, bringing her cubs to a point near the kill, but then lying down and waiting for the cubs to discover the prize. Frequently this takes them a long time as they have to learn to smell out the decaying meat and to observe the insects that soon discover even a well-buried carcass.

By ten weeks of age, the cubs make regular hunting trips with the tigress, chasing monkeys and birds. When the cubs are six months old they lose their milk teeth and grow a second set of teeth. These are hollow, and the permanent teeth grow inside them until they are ready to shed the hollow set. When their teeth are strong enough to produce a powerful and damaging bite, the tigress takes her brood to hunt larger prey. She brings down a deer by hamstringing it so that it cannot escape, and then she withdraws from the hunt, leaving the injured prey to the cubs. With the first few victims the cubs take a long time to kill the unfortunate animals. Efficient killing seems to be a matter of training and experience rather than instinct. However, the cubs gradually learn the skills of neck biting and to throttle and sever the vertebrae.

As their experience grows, the cubs take an increasingly active part in the hunts. The tigress is usually on hand in case the prey turns out to be too strong for the young. If they make stupid mistakes or are excessively playful and noisy during the stalk, she clouts them. It is at about this time in their development that young tigers, in the process of learning, tend to kill more than they need for food. This has led to an unfair reputation for destructive savagery and of killing for the fun of it — a reputation that, with fairness, can only be attributed to man.

As the cubs grow older, their mother leaves them for long periods. The father sometimes stays near the tigress after mating and has been known to form a family group with them for a year or more, but this is uncommon; males tend to be more nomadic than tigresses. By the time they reach about three years of age, the young tigers are ready to tackle big game and will probably kill their first buffalo. The tigers are about 2.25m in length by this time, agile and strong, but still showing their rough, white-spotted immature coats. During the following year they achieve their full status as adults.

Food

A glance at the structure of the tiger's jaws and the characteristics of its teeth will at once show an observer what kind of diet the tiger enjoys. It eats large quantities of meat, which

---

Although the male tiger seldom helps the female to rear the young cubs, it sometimes goes hunting with the the mother to supply their offspring with food. Deer are one of the main sources of food for the tiger family. When they are about seven months old, the cubs start to kill prey themselves although they remain with the mother until they are two years of age, being trained in the arts of hunting and stalking prey. During this period, the male tiger leaves its mate to hunt and live on its own.

swallows in large lumps without careful chewing. The tiger is not too particular about the condition of its meat. It will eat fresh kills or carcasses which it has hidden for some days, and even carrion that is in an advanced state of decomposition.

Most of the tiger's prey is middle-sized but it will also take small creatures, such as rats, frogs and even insects if larger prey is scarce. Even strong tigers seem to prefer to avoid attacking large game such as the gaur, but hunger will drive them to risk being gored.

In seeking out their prey tigers cover territories of various sizes. Siberian tigers are the most nomadic and have been known to travel as much as 1000km in 28 days in search of new hunting grounds. Bengal tigers, living in territories rich in game, concentrate on relatively small areas of jungle. The male will patrol about 55 square kilometres and the female will concentrate on a slightly smaller one of about 45 square kilometres.

These hunting grounds are not defended rigidly; a male will let other males through its territory without challenge but will attack if the strangers show signs of settling down or mating a tigress in the area. Females do not defend their territories although they mark them out with scratch markings on trees and the ground and exposed droppings.

The tiger usually eats large quantities of meat and then may eat almost nothing for several days before killing again. A grown tiger will eat about three tonnes of meat a year — this amounts to about 30 domestic cattle or 70 deer annually. A tigress and her two or three cubs may eat 280kg of meat every 20 days, but the training of the cubs may involve the killing of more game than the figure represents.

The order in which the tiger eats his kill varies but generally he starts by eating the beast's neck, then flanks, belly and the soft parts of the back. He will eat a buffalo in three or four days and a sambar deer in a couple of days, consuming up to 30kg at a meal. The eating process is frequently interrupted for copious drinks of water.

Enemies

Tigers have few natural enemies. The only one that has confined the animal to restricted areas of the jungle and taiga is his most powerful and dangerous enemy — man armed with a high-powered rifle.

The tiger has an unaccountable passion for porcupine meat, a taste which has indirectly killed many fine tigers. The tiger kills the porcupine by hitting it with a paw and once the needles penetrate deep into the animal's sensitive pads they are extremely hard to draw out. The tiger, maddened with pain, often bites at its victim and even swallows chunks of the porcupine with the quills still attached. The quills gradually move into the tiger's internal organs and cause serious damage. A quill or two in a tiger's jaw can make it impossible for the animal to eat anything but the most tender meat, and even that will hurt. Such tigers will kill on a huge scale, eating only the softest parts before abandoning the kill.

Jungle tigers eat wild boar in large numbers but the boar is a powerful and dangerous animal. The thick shield of cartilage covering the ridge of muscle on its back makes it difficult for even a tiger to penetrate to the backbone. There are several instances reported of boars disembowelling tigers which have had the temerity to attack them. The boar is a durable fighter, not cowed by severe wounds and always alert for the slightest advantage that will allow him to slash with his curving tusks.

The huge gaur and even the domestic buffalo have been known to turn on an attacking tiger and succeed in goring it. The tiger seems to be aware of this danger and often hamstrings the animal so that it cannot manoeuvre.

Langur monkeys frequently disturb a tiger's stalk by calling warning sounds which carry for quite some distance in the forest, so making life hard for the hunter. Small barking deer, too, will alert a tiger's prey to its danger, as will ever-watchful peacocks.

Camouflage

The brilliant orange-red of the tiger's coat, with its black stripes, is easily seen by the hunters when the animal moves against a background of bright green jungle plants. Even when the animal is stationary, trained eyes will pick it out without much difficulty. In the burnt-out colours of the hot season it is harder to see the tiger, but its camouflage is certainly not of the conventional kind. In any case, most animals are not very good at picking out stationary objects, but they are quick to spot movement of even a subtle kind.

Many of the tiger's prey animals are colour-blind and see only in tones of grey, so the eye-catching background colour of the tiger's coat is lost. The stripe pattern seems to arouse the curiosity of the prey animals, which may have difficulty in perceiving the tiger's outline but see the strange movement of vertical stripes (like black shadows) against the shadows of the foliage. By night the tiger's stripes may have the effect of a disruptive pattern, once more making the tiger's shape hard to define.

Maneaters

Tigers live their lives dangerously and frequently sustain injuries. They generally recover quickly even when the damage is quite severe, but some injuries are disabling for long periods or even permanently. Damage to paws, limbs, teeth and jaws can make it impossible for a tiger to hunt effectively. When this happens the tiger may turn to easy prey, possibly taking a large proportion of domestic animals or, as a last resort, killing and eating humans.

It is odd that so vulnerable a prey as a human being is not more attractive to tigers, but they seem to avoid contact with men as far as possible. If only soft-skinned, slow-moving prey can be caught, then a weakened or aging tiger soon forms a taste for people. Once this happens, the tiger may enter villages and even houses in search of this prey. A tiger that has turned maneater can close down the activities of an area from dusk to dawn as villagers fear to leave their houses.

Maneaters seem to develop skills that enable them to avoid many of the traps laid for them, and hunters are usually called in to shoot the animals. The most terrible of the maneaters was probably the tigress shot by J.Corbett in the Champawat district of India in 1907. This tigress is believed to have killed over 430 people.

Anatomy

The tiger's body is structured for strength and agility. The short muzzle makes a tough, durable seat for the immense canine teeth. The long, curved spine is supple while the tail helps the animal's balance. Its short neck and muscular forepaws give the tiger great leverage in tearing and twisting the bodies of prey animals, the retractile claws being unsheathed to sink about 125 mm deep into a victim, and the long hindlegs giving the tiger leverage for its astonishingly long springs.

The tiger can produce bursts of speed over short distances and pin down most prey, but it cannot maintain its speed for more than 100m or so. Its skeletal and muscular structure enables the tiger to leap after its prey for five or six metres, clear heights of up to two metres with ease, and drop on to prey from heights of as much as 10m.

The head

The tiger's skull is well equipped for its life as a hunter. Its stabbing canine teeth, about 75mm long in the upper jaw and some 65mm in the lower jaw (one exceptional tiger had upper canines measuring nearly 140mm),are the tiger's principal armaments for killing its prey. The tiger uses its canines also to shear off large chunks of meat although it can cut through meat with its molars as well. The incisors, which are quite small, are used to scrape, gnaw and nibble.

The tiger's jaws and razor-sharp teeth are its most deadly weapons. Viewed from above, the bony crests of the skull, to which the powerful jaw muscles are attached, are visible. The sturdy cheek arches support the muscles used in biting and there are 28 or 30 teeth.

At a glance
Bengal tiger
Panthera tigris tigris

Class	Mammalia
Order	Carnivora
Family	Felidae
Genus	Panthera
Length, head to rump	1.80 – 2.80m
tail	80 – 94cm
Height at the shoulder	90cm
Weight	227 – 272kg (The female is a little smaller than the male.)
Gestation period	92 – 112 days
Litter size	2 – 4 cubs, exceptionally up to 7.
Weight of cub at birth	800 – 1500g
Length of cub at birth	31.5 – 40cm
Longevity in the wild	15 – 25 years
Diet	Carnivorous

The arrangement of the tiger's jaws allows hardly any lateral movement. This means that the animal's bite is extremely strong and reduces any danger of the jaws becoming dislocated by the struggles of a large prey while the tiger grips it. The jaw structure also makes it impossible for the tiger to masticate food fully, so it tears off lumps of meat and swallows them whole.

The tiger's tongue is long and covered with backwards sloping papillae (horny projections). The animal uses them to scrape tissue from the bones of its victims and to lick its own coat clean. The tiger's taste organs are minute, which may suggest that its sense of taste is not very highly developed.

The eyes are protected by strong orbits of bone, the pupils being vertical slits which expand to about 20 times their narrowest width. They are narrowest in daylight and widest at night, when the retina must collect as much light as possible to give the tiger good night vision. The retina is highly sensitive and is assisted by a 'tapetum' layer which lies behind it, reflecting light back through the retina again so that the light strikes the retina twice. It is this layer that gives the tiger's eyes their reflective character, making them seem to glow like live coals in the dark. Despite the fact that tigers have such good night vision and can see in depth, they are colour-blind, seeing images in tones of grey.

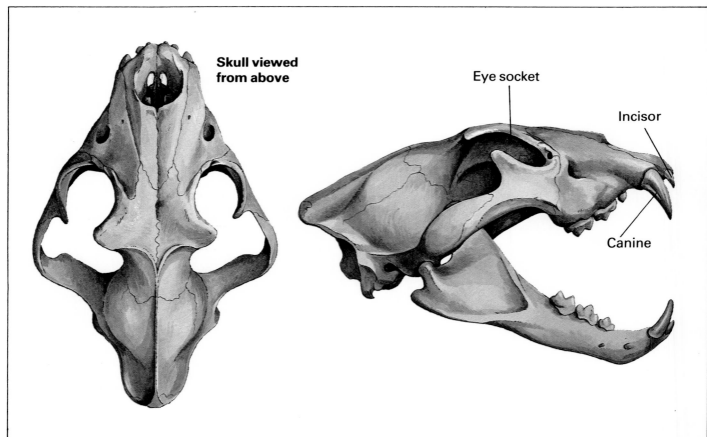

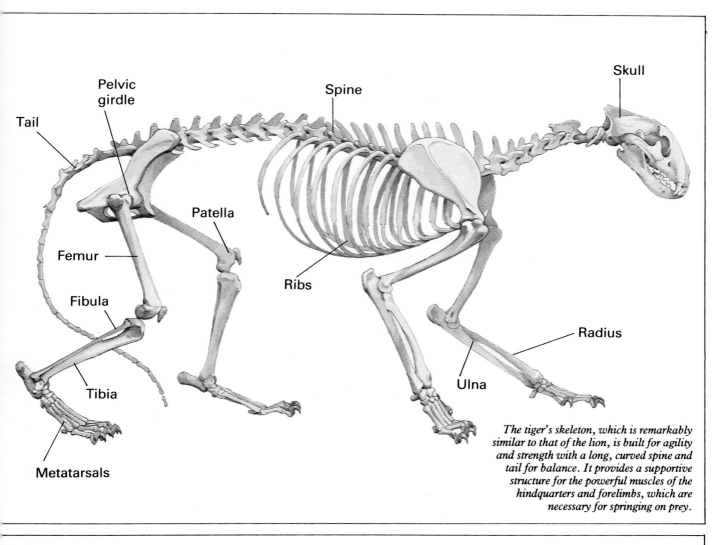

The tiger's skeleton, which is remarkably similar to that of the lion, is built for agility and strength with a long, curved spine and tail for balance. It provides a supportive structure for the powerful muscles of the hindquarters and forelimbs, which are necessary for springing on prey.

On the move

The tiger walks gracefully and silently. The enormous development of muscle around the shoulders and in the forelimbs gives it an undulating walk, with a stride covering about 1m at a time. Like all cats, the tiger walks in a digitigrade manner, its toes and the balls of its feet touching the ground while the palms and soles are raised clear.

The tiger makes tracks or 'pug marks', about 150mm in diameter, with its feet. The tigress's pug marks are more pointed than those of the male. The animals move on softly-cushioned paws that are ideal for quiet stalking in the jungle and grassland where they hunt. Their tracks show five toes at the front, and a hind footprint of four toes, the fifth toe being held clear of the ground.

Generally the tiger has its claws retracted, but when ready to kill or to defend itself it protrudes the claws ready for action. The claws are attached to the terminal phalanxes and when the tendon of the protractor muscle is drawn back, the terminal phalanxes pivot to bring the claws forwards. Afterwards, when the claws are no longer required, the reactor ligament contracts, the protractor muscle relaxes, and the terminal phalanxes pivot the claws back up into their original sheathed position as shown in these diagrams of the claws.

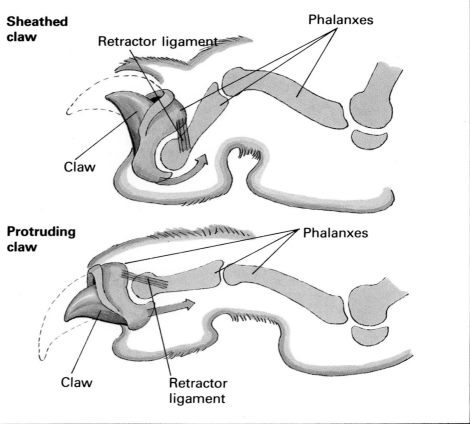

The Sidewinder

All day long the cicadas had sung in chorus with a choir of insects and now that the sun slipped below the purple horizon their electric chirruping died away. The furious heat of the afternoon had, for a while, nearly silenced them. The whole desert, its sand and baking stones and the living creatures that knew the secrets of survival there, shrank and quivered under the irresistible glare of the white sun. Slowly the insects made again the blanket of sound that oppressed the desert day in and day out, except for those moments when even they acknowledged the power of the sun.

Now that the shadows streamed across the desert slopes, rippling along its sides from clusters of triangular, sand-blasted pebbles, beetles made busy tracks in the sand. A stunted yucca tree stood with its angular arms akimbo, branching away from the prevailing wind. Its foot clutched into a stony patch of ground which seethed with termites, and nestling under a tuft of the dusty, spiky leaves was a small cactus wren, which was watching a small hole beside a dead-looking creosote bush. The shadow that filled the mouth of the hole shifted as a triangular sandy head emerged slowly into the dim light. Before the snake moved more than a few centimetres forward it paused, raised its head a little, opened its mouth and flicked its dark tongue out from between narrow lips.

The forked tongue licked the delicate messages from the air and ground and drew them back into the mouth. The Jacobson's organ above the snake's palate responded to the minute traces of chemicals on the tongue. None of them indicated the presence of an enemy, so the snake decided that it was safe to advance from its cool shelter.

The female snake slid slowly down the gentle slope. Her sandy-coloured body was clearly marked with a pattern of darker diamond shapes; when moving she was easy to see but when she lay still it was hard to pick out the plump coil of sandy brown against the desert's sand and stones.

With many pauses and much flicking of her tongue, testing and retesting her environment she made her way to the shelter of a neighbouring creosote bush. She seemed to peer at the area around the straggling, dried-up stems, her lidless eyes moving minutely further into their sockets as she tried to focus on the terrain. (The spherical lenses of her eyes had no muscles to modify their curvature so their focus was adjusted by moving them backwards and forwards in her head.) She tilted her head a fraction to shelter her eyes from the last rays of the sun. Shade from the pronounced brow ridges over her eyes gave the comfort her eyes needed, and the vertical slits of her pupils closed a degree to filter out the last scintilla of light that shot horizontally from the sinking sun.

The sidewinder's examination of the ground around the bush seemed to satisfy her. She curled up carefully in the dip near the foot of the bush, flickered her tongue in several directions, and composed herself for a wait.

It was not far from this spot, three months ago, that she had detected a male sidewinder, his head held high, following her scent, steadily angling his way towards her. He was within range of her poor vision when another male crossed the sand to intercept him. She watched both snakes rear up side by side. Their lower bodies locked together and their upper bodies wrestled. They were like two humans with their forearms locked in an Indian wrestling match.

Opposite page:
The sidewinder derives its name from its curious sideways locomotion. It uses this odd method of movement to give it an effective grip in the loose sand in which it lives in the deserts of south-western North America. Thus it leaves characteristic tracks in the sand resembling a series of 'J's, or parallel waving lines which are hooked at one end. It moves like a small coiled spring and only two parts of its body are in contact with the ground. These remain stationary while the raised parts move lifting half the body to a completely new position.

These little prairie rattlesnakes are found across the western great plains of the United States and are universally feared. Although they do not normally kill humans if bitten, they can cause death in untreated cattle and horses which tread on them. About 60cm in length, they are greyish-green in colour with dark blotches running along the back and characteristic small scales on top of the head. During the cold winters they live in rocky dens below the frost-line.

They swayed gracefully, bending one another's heads towards the ground, recovering, and pressing again for a better purchase. The struggle went on for several minutes before the first male succeeded in pressing the other to the ground. But the deadly fangs had not even been bared and neither snake had struck at the other.

The stronger snake approached the female obliquely. He could tell from her scent that she was ready to mate with him. At two and a half years old, she was receptive to males for the first time. She had often seen them wrestling, sometimes to prepare for mating.

She lay still as the male gently stroked her glossy coils with his chin. He licked at her neck and wound his body round hers, first one way and then another. The scutes under his belly, just in front of the narrowing of his tail, swelled outwards a little, the hemipenis emerged and he pressed it into the female. For hour after hour the pair lay entwined. Occasionally a ripple of convulsive movement ran up the male's body, but the female lay supine. It was about 14 hours later that she seemed to tire of the situation and began to pull away from the male. She dragged him after her for some metres before he withdrew and released her.

This had all occured in the late spring and now she carried in her body seven young snakes, each in its delicate membrane. Eventually she would give birth and the live young would break free of the membrane sacs, ready to fend for themselves at once. She would keep them near her for about eight days until they sloughed their skins for the first time, and then they would leave her to feed as rapidly as possible so that they would withstand the winter cold. Many would die in the fierce sun or from predatory king snakes and roadrunner birds. (The king snakes were almost immune to the

Sidewinders are most active during the early part of the night while the air is cool. Because of their unique infra-red heat receptors, they need not hunt in daylight and can locate their prey in complete darkness. In the heat of the day, they coil themselves up like miniature springs and, in a series of looping movements, they bury themselves in loose sand with only the eyes exposed. They usually choose a shady spot below a yucca tree or a desert creosote bush.

sidewinder's poison, and the roadrunner was so fast that it could afford to tempt the snake into strikes that just missed their target until the venom supply ran out and the bird could strike back with its powerful beak.) More of the young sidewinders would die before they could find a suitable cave or deep hole in which to hibernate for the winter. There was never any difficulty over territorial defence for the sidewinders. There was enough of the harsh land to go round after nature had thinned their numbers.

The desert was now as dark as it ever was in the summer. The light of a huge moon turned the landscape into a complex pattern of silver and black, and the sky blazed with millions of points of vibrant light. The sidewinder lay still. She had chosen her place with care. She had noticed signs of a trail, barely scuffed into the desert sand. Her tongue had picked up the confirmation from the air and she had lain patiently in wait for the rat she believed might come back along the trail. No sounds disturbed her watch for she could not hear. The owl's cry went unnoticed but a faint vibration on the ground caught her attention. Footsteps scampered across the rat's trail several metres away as one ground squirrel chased another towards their burrow. The vibrations ran through the snake's sensitive skin and she raised her head to test the air. Her eyes caught a movement some metres away, but her weak vision could not sort out the image: she was simply aware of a shape. She turned her head towards it. Warmth from a living creature was detected by the astonishingly sensitive pits between her eyes and nostrils. The signals were faint but clear and on the move. Gradually the heat image resolved into something that had both direction and distance. Soon she knew that it was the rodent she had waited for.

The timber or banded rattlesnake, another relative of the sidewinder, lives in the eastern states of the United States and is easily recognisable by the dark chevrons on its back. Rattlesnakes are best known for the rattle in their tails which vibrates, making a buzzing, whirring sound when the snake is disturbed. The rattle consists of several interlocking shells which were originally the last scale on the tail tip. Every time the snake moults and sheds its old skin a new shell is added to the rattle which rarely exceeds 14 segments in a rattlesnake living in the wild.

Her tongue flickered about her lips but all else was still. The kangaroo rat hopped nearer; he was wary but ignorant of the ambush laid for him. The 'heat image' told the snake that her prey was nearly in range, although, in the shadow, the cautious creature was invisible to her eyes. Slowly, very slowly, the sidewinder's head rose upwards and leaned back. The movement was imperceptible.

The rat hesitated half the snake's length away. He felt there was a threat, but where? There was a moment of panic and then he knew. The sidewinder struck so fast that she almost left the ground. Her mouth stretched wide and her fangs advanced into their attacking position as she thrust her head through the air to drive them into the rat's flank. The kangaroo rat screamed and leaped sideways wrenching himself free. One of the sidewinder's fangs snapped off and remained in the rat.

The rodent scampered a few paces and stopped. It seemed as though he could not decide which way to go. He shivered, put forward a foot, overbalanced and fell onto his side. The venom that had flowed through the fangs into his bloodstream was already breaking down the tissues of his organs and causing massive internal haemorrhaging. The sidewinder crept closer but waited for the twitching of the rat's body to cease before flicking it with her tongue. The rat was dead and the entire kill had taken less than 40 seconds from strike to death. It might have taken longer had the fangs only scratched the rat, instead of puncturing a blood vessel so that the venom flowed straight into the bloodstream.

The sidewinder lay beside her pray, licking it over. She moistened its fur until the corpse was slippery; then positioning herself at its head, she opened her mouth wide and began to work the rat down her throat. The snake's back teeth were little backward-pointing thorns. They secured a grip on the rat and then the flexible jaws dragged the meal a little further into the throat. Several times the sidewinder paused to rest, then the front half-jaw reached forwards to drag the rat inwards until it had passed the narrowest part of the gullet. Once this point had been reached, the muscles along the snake's sides squeezed the prey down into its stomach.

The sidewinder lay still, waiting for the digestion of the rat to proceed. It was some hours later that the sky paled to meet the dawn, and the warmth of the sun revived the dormant snake. She still made no move but waited until her temperature was about 35°C before moving to the shelter of a hole.

The satisfied snake spent the morning alternately sunning herself on the hot sand and then making her way by the curious zig-zag heel-and-neck gait back to the shade of her hole. In this way she kept her temperature at a comfortable level.

As she moved, her skin shone in the hard bright light. It was new skin, not more than a month old. The change had happened after several days of discomfort. The sidewinder had been irritable, and no wonder. Her skin looked dull and dry and her eyes had turned a milky white. (The snake's skin covered the eyes with a transparent film, rather like a pair of spectacles. This normally transparent film of skin protected the lidless eyes from dust and the abrasions of windblown sand. When the dead layer of skin that covered the rest of her body was ready to slough off, the film that covered her eyes would fall away with it.)

One morning she had blindly rubbed her lips against a piece of rock, fraying the skin away. She scratched the base of her tail, just where the lowest section of her rattle joined her body, and little by

The sidewinder is often known as the horned rattlesnake because of the two raised head scales above its eyes which resemble horn-like growths. These provide the useful function of protecting the snake's eyes from loose sand particles when it is buried in sand during the day. Running backwards from each eye across the broad head is a dark stripe. The pale grey or light brown body blends beautifully into the desert background to provide an effective means of camouflage.

little the old skin peeled away. The scaly skin fell from her eyes and she saw clearly once more. For a while she warmed her newly revealed skin in the sun before seeking the concealment of a ground squirrel's hole. Her tail had acquired another rattle in the shedding of her skin and it now sported six.

Still heavily rounded from her meal, the sidewinder 'stepped' her way, tail to neck, towards the apron of sand to sun herself. She did not hurry but her movement caught a pair of sharp brown eyes. Equally sharp ears pricked, a russet-coated coyote rose slowly from where he had crouched in the shade of a yucca tree. He waited until the snake had warmed herself for several minutes and began to wind her way back to the shelter of her adopted hole before he trotted briskly towards her. Heat-sensitive pits, the Jacobson's organ and eyes told the snake that she was in danger. She rapidly coiled and her head craned up and back in an 'S' bend, ready to strike. Her tail rose slightly and her rattle whirred. The sound was like seeds being shaken in a gourd. Her body swelled out a little and breath hissed from a gaping mouth. The one curved fang displayed was fragile and dangerous; the other was no more than a venomous stump but still capable of infecting the coyote with her poison.

Although it is deaf, the sidewinder can react quickly to movements and vibrations around it. Its eyes, in particular, are very sensitive to small movements in the snake's vicinity. In this way, it can sense the position of its prey and strike. Its diet consists of small rodents, especially deer-mice, kangaroo rats and spiny pocket mice, lizards, small birds and even other snakes.

This young rattlesnake has been fortunate to survive in the harsh conditions imposed by life in the desert. Although rattlesnakes give birth to litters of between 10 and 20 live young, few reach adulthood, many falling prey to such predators as hawks, skunks and snake-eating snakes. Other hazards include the intense heat of the sun, the bitterly cold nights and starvation. Rattlesnakes usually mate in the spring and it is not unusual for a female to give birth to a litter as much as two years later in some instances. The number of litters per year is dependent on the climate.

The coyote snapped at her from a safe distance and circled round her. She writhed round to face her attacker. The coyote was in no hurry and stepped back. The sidewinder began to feel the heat of the sun which was near its height. She knew she must find shelter soon or dry out and die. She made a lunging break towards the cooler entrance of the hole but the coyote crossed her track, snapping his sharp white teeth near her neck and causing her to coil back in defence. She was becoming confused with the heat and the constant circling threat of the four-legged, yelping creature that danced about her every time she tried to escape. She writhed with desperation.

The commotion caused two ground squirrels to pop from their holes nearby, chattering with alarm and curiosity. It was only a moment's distraction, but enough. The coyote spun round to face the noisy pair of creatures, and the sidewinder made a last, lunging move to the tilted piece of rock that sheltered the mouth of her hole – safety was so close. The coyote realised his mistake and flung himself after her but missed his chance. With a final skirl of her rattle she disappeared deep into the narrow tunnel. The coyote dug furiously for her, but when he heard her hiss in the darkness of the hole he thought better of it. If he had succeeded in keeping her out in the sun for another few minutes, she would have become helpless with heatstroke and he might safely have eaten her.

The sidewinder cooled off in the darkness and made her way out through another passage to watch the sun moving down towards the horizon. She was well nourished from her last meal so she lay quietly at the hole's entrance, occasionally testing with her forked tongue the savour of the changing scene about her.

Anatomy of the snakes

The lack of limbs and the smoothness of the body contours make the snake's passage through undergrowth silent and hard to detect. Speed in the strike, sensitivity to the environment and adjustment to living in trees, water and desert sands, as well as jungles and grasslands, make snakes a highly successful sub-order of reptiles.

The evolution of the snake's long, slender body has brought a number of changes in the shape and arrangement of its internal organs. The intestines, which in mammals are coiled, form a long straight tube, the forward extension of which is the stomach. The problem of fitting in the twinned organs, such as the bi-lobed liver, the lungs, kidneys and genital glands, is solved by staggering their position in the body, and developing one of each of the organs fully while leaving the other twin small, almost vestigial.

The snake's brain is long and narrow. Unlike that of a man, it does not occupy the greatest width of its head. The greatest width is at the wide point of the extremely flexible jaws. The smallness of the snake's brain is not an indication of a weak intelligence, but of the demands of its lifestyle: it has no need for speech centres, auditory centres, nor does it need any brain space for the complex co-ordination of the limbs that it does not possess. The proportionately great length of the animal's spinal cord supplies the impulses and responses that in other animals might be produced by the brain.

Rubbing the whole length of its body on the ground produces a heavy degree of wear on the snake's skin. Like most reptiles, snakes have well-armoured, tough hides. The skin's deepest layer is fibrous and contains the pigment cells that give the snake its characteristic markings and colours. (It is this layer that gives snakeskin its leathery quality.) Above it lies a layer of skin which constantly renews the horny material of the outer layer, which is called keratin. It is an inflexible substance, and the snake can move without breaking its skin only because of the complex pleating of the skin.

The clear scales of keratin are each about 0.001 mm thick. The snake has broader areas of keratin in the form of scutes, or plates, under its belly, where the wear and tear is greatest and where frictional resistance to its passage across the ground must be least. The polished scutes slip easily along any natural surface the snake is likely to encounter.

At a glance	
Mojave Desert sidewinder	
Crotalus cerastes cerastes	
Phylum	Chordata
Class	Reptilia
Order	Squamata
Family	Crotalidae
Length	male c.54cm (exceptionally up to 59cm) female c.53.5cm
Length at birth	c.17.5cm
Sexual maturity	Attained at 2½ years. Female first gives birth at 3 years.
Gestation	140 – 200 days
Average number of young per birth	Nine in litter
Birth	Live (not in eggs, requiring incubation).
Diet	Small reptiles, birds and mammals.

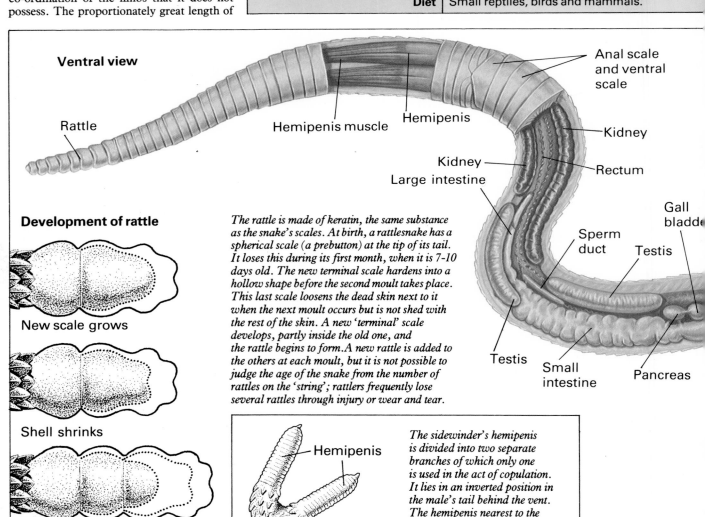

Ventral view — Rattle, Hemipenis muscle, Hemipenis, Anal scale and ventral scale, Kidney, Kidney, Rectum, Large intestine, Gall bladder, Sperm duct, Testis, Testis, Small intestine, Pancreas

Development of rattle
New scale grows
Shell shrinks
New shell grows

The rattle is made of keratin, the same substance as the snake's scales. At birth, a rattlesnake has a spherical scale (a prebutton) at the tip of its tail. It loses this during its first month, when it is 7-10 days old. The new terminal scale hardens into a hollow shape before the second moult takes place. This last scale loosens the dead skin next to it when the next moult occurs but is not shed with the rest of the skin. A new 'terminal' scale develops, partly inside the old one, and the rattle begins to form. A new rattle is added to the others at each moult, but it is not possible to judge the age of the snake from the number of rattles on the 'string'; rattlers frequently lose several rattles through injury or wear and tear.

Hemipenis — *The sidewinder's hemipenis is divided into two separate branches of which only one is used in the act of copulation. It lies in an inverted position in the male's tail behind the vent. The hemipenis nearest to the female is used.*

Because they have poor eyesight and are completely deaf, snakes compensate for these deficiencies by extra acuteness in other senses. They are extremely responsive to vibrations in the air about them and within their own bodies. Their skin is sensitive to earthborne vibrations. The snake has hardly any taste buds and depends on the responses of the chemico-receptor surfaces of the Jacobson's organ. From here, the messages are transmitted along nerves that lie close to, but separate from, the olfactory nerves to the brain. The pit vipers, including the rattlesnakes, have an infra-red heat receptor. Each pit has an inner and outer chamber divided by a 1/100mm-thick membrane which acts as a receiver for the infra-red signal. The sensitivity of the organ is remarkable. It is swifter to react than any known made heat sensor. It will detect temperature differences as low as 0.003°C. In the pit viper family, the heat sensor lies in a pit that is situated between the nostril and the eye. The position of the pits on either side of the sidewinder's head allow the snake to percieve its prey stereoscopically. Each sensor operates through a horizontal arc of about 100°, each arc extending some 10° past a centre line. The arcs overlap to give 'binocular vision'. The sensitivity of the sensor is such that the sidewinder can locate and strike its prey in total darkness. Thus the rat is 'seen' as a warmth pattern by infra-red impulses.

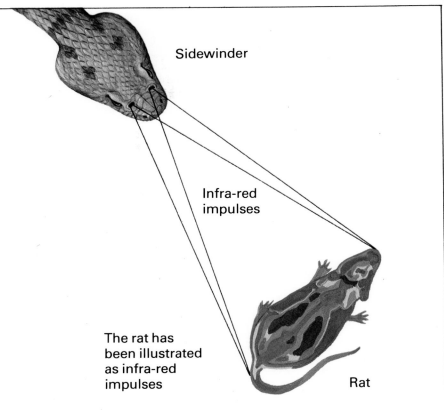

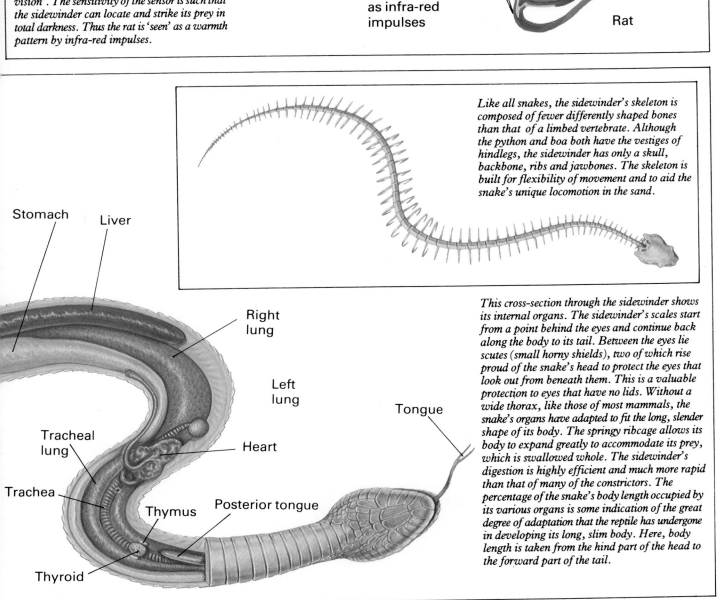

Like all snakes, the sidewinder's skeleton is composed of fewer differently shaped bones than that of a limbed vertebrate. Although the python and boa both have the vestiges of hindlegs, the sidewinder has only a skull, backbone, ribs and jawbones. The skeleton is built for flexibility of movement and to aid the snake's unique locomotion in the sand.

This cross-section through the sidewinder shows its internal organs. The sidewinder's scales start from a point behind the eyes and continue back along the body to its tail. Between the eyes lie scutes (small horny shields), two of which rise proud of the snake's head to protect the eyes that look out from beneath them. This is a valuable protection to eyes that have no lids. Without a wide thorax, like those of most mammals, the snake's organs have adapted to fit the long, slender shape of its body. The springy ribcage allows its body to expand greatly to accommodate its prey, which is swallowed whole. The sidewinder's digestion is highly efficient and much more rapid than that of many of the constrictors. The percentage of the snake's body length occupied by its various organs is some indication of the great degree of adaptation that the reptile has undergone in developing its long, slim body. Here, body length is taken from the hind part of the head to the forward part of the tail.

Rattlesnakes

The family *Crotalidae* contains six genera of snakes, but only two of these, the *Crotalus* and *Sistrurus* genera, are true rattlesnakes. All members of the other genera are pit vipers but they have no rattles. Rattlesnakes, of which there are 30 species and about 65 sub-species, occupy various habitats but are found only in the New World. The *Sistrurus* genus are pygmy rattlers, the largest of which is the eastern Massauga rattlesnake (*S. catenatus catenatus*) which grows to about 95mm. The largest of the rattlers is the eastern diamondback rattlesnake (*C.adamanteus*) which grows to about 2.5m, weighs up to 10kg, and like most rattlesnakes, is thick-bodied. It produces extremely large amounts of venom; 'milking' the eastern diamondback (by making the snake bite on the lip of a vessel to eject the venom for the preparation of antivenin), herpetologists have obtained 1,050mg of the deadly venom from only one snake at a single session.

The sidewinders comprise three species of the *Crotalus* genera and all are found in the south-west of the North American continent. Many of the world's desert-living snakes adopt the method of locomotion that gives the sidewinder its name, but here that name is applied only to the three snakes of the *C.cerastes* species. Sidewinders are all small but they move more quickly than most snakes.

Mojave Desert sidewinder (*Crotalus cerastes cerastes*). See *At a glance* panel.

Colorado Desert sidewinder (*C.c. laterorepens*) is the largest of the sidewinders. Males average a little over 540mm in length and females about 562mm, but exceptional specimens may reach 767mm.

Sonoran Desert sidewinder (*C.c. cercobombus*) is the middle-sized sidewinder. The largest grow to a length of about 628mm.

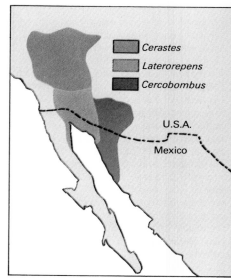

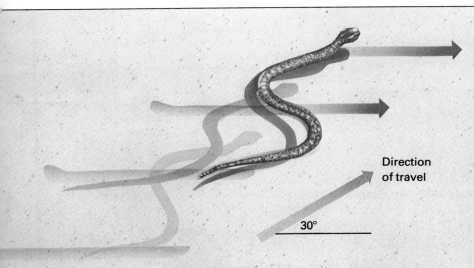

Locomotion
Snakes find it difficult to move over shifting sand, but the sidewinder has a special adaptation of normal snake movement. Instead of trying to thrust its whole length across a surface that offers little purchase, the sidewinder uses temporary points of purchase by touching the ground with only two, comparatively small areas of its body at a time. Its gait resembles diagonal stepping. The Mojave Desert sidewinder is probably the fastest of the sidewinding snakes. It can travel at about 3-4 km/h. The tracks it leaves are an odd-looking series of 'J' shapes which lie at an angle of 30° to the direction of travel. This diagram shows how the snake makes its bizarre looking tracks. Only two sections are in direct contact with the ground, the rest being held clear of the surface.

Fangs and venom

The venom of the sidewinder is poisonous enough to kill a small mammal or reptile quickly. The venom enters the animal's bloodstream through the snakes's fangs during the brief strike.

The colour of the venom varies from pale straw to orange, and is secreted by the glands near the outer edges of the upper jaws. Some of the muscles that operate the jaws also eject the venom from the glands into the fangs. The venom then flows down the hollow fangs and into the snake's prey. There, the venom's complex enzymes and proteins attack the walls of the capillaries and destroy the red corpuscles of the blood and the connective tissues of the organs. This destruction not only kills the victim, but also aids the snake's subsequent digestion.

To be effective, the venom must enter the bloodstream of the victim. An animal can swallow the poison without ill effect as long as there is no internal injury that would allow the venom to escape into the bloodstream. Many protein substances in the normal human diet, egg white for example, would be poisonous if they entered the blood.

The fangs themselves are hollow and rather like hypodermic needles. They are delicate instruments which frequently break and are sometimes left in their prey. The snake grows replacement fangs between two and four times a year. In order to be accommodated in the mouth, they are swivelled backwards to lie parallel to the upper jaw. When the snake opens its mouth the fangs may remain in the 'safety' position and can be extended ready for use. The fangs grow from the maxillary bone, which is pivoted forwards by means of muscles, bones and tendons in the upper jaw.

The snake eats its prey by swallowing it whole. The great elasticity of the jaws enables it to swallow surprisingly large prey. In some cases, the snake protrudes its glottis if a meal fills its throat so tightly that it might otherwise be unable to breathe. The diagrams show the snake's skull, including a lateral view of the skull in striking position. The fangs are set inside special maxillary sockets ready to strike their victim.

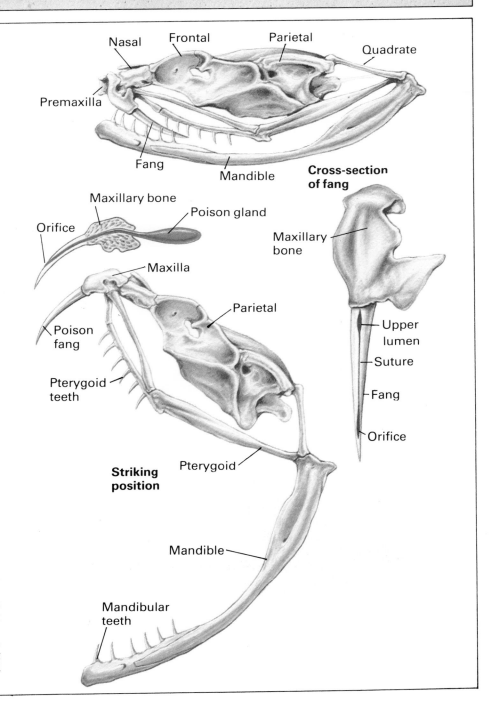

The Lemur

Brittle flakes of kily pod filtered down through the branches, tapping and glancing from the glossy vines that hung from the tall tree's lower branches to the tangled undergrowth below. The forest was still, just before the dawn light bleached the starlit sky. A pause in the scurryings and night noises quietened the cool air before the beginning of the new day; except that high in the kily tree's branches something was busy shredding old pods. Then even this fidgeting stopped as the faint light of the sun paled the eastern sky.

A wide river with sandy-red banks curved through the forest of kily trees (tamarinds), whose bushy round tops swept upwards for 30m. Their branches and oak-like leaves threw a dappled shade as the sun climbed higher over the horizon and the dark leaves of a dense fall of vines, tumbling and twisting their way from tree-top to ground, curtained the fringe of a small glade.

Two-thirds of the way up the tallest tree, a small animal moved along a branch and began picking at the fur of a larger companion. Their black-and-white ringed tails waved slightly in the light and shade of the leaves as the young lemur tidied the tangles from her mother's clear, light grey coat. Delicately, she worked a burr clear of the fine fur under her mother's neck, using her teeth as a scraper to work the irritant free. The offspring presented an arm to her mother, and the older female began to groom it meticulously. She concentrated hard and soon the young lemur's fur sparkled.

The ringed tails of the troop flashed in the early sunlight as the animals warmed themselves. There were 14 animals in the branches of two large trees. The three fully-grown females sat on a couple of branches some metres away from the eight males, who were scattered about several levels of the trees. Three juvenile females were fairly close to the other females. A young male watched the troop from about 30m away. He avoided the males of the troop who, during this second week of April, seemed to have lost their previous tolerance of his presence. He nursed a raw slash in his upper arm where, the day before, another male had inflicted a 5cm-long wound with his canine teeth.

He moved a few steps towards the females, but a chorus of clicks and grunts greeted his move at once and he dropped to the ground to sniff among the fallen debris from the trees. A large male with wide, dark eyerings cuffed away another male which squatted on a branch between him and the group of females. He strutted aggressively towards the females but they barked shrilly at him and leaped to another tree. All three flew through the air in rapid succession, turning in mid-air to lean backwards before they landed. At the moment of impact, they grabbed for handholds and footholds and then scampered up to the large horizontal branches. The male flung himself across the gap and chased a female who was ready to mate.

Her genitals were flushed a bright pink, making her receptive condition clear to the rest of the troop. She paused on a wide branch and marked with her scent the point where the branch emerged from the main trunk of the tree. She climbed higher as the male paused to mark the same spot with his own scent, rubbing one palm across a nipple-like gland which projected from a point high on his chest. A greasy liquid smeared on to his palm, and he quickly wiped it on to the spot the female had marked with her rump. He followed her with a curiously jaunty, stiff-legged walk up onto a broad bough.

40

Ring-tailed lemurs are naturally gregarious, living in large social groups, or troops, of up to 24 members. They travel, eat and sleep together, defending their territory fiercely against attacks and intrusions by other troops. At night, they settle down to sleep in tall 'sleeping' trees which are regularly used for this purpose. They have an elaborate system of communication consisting of a variety of howls, grunts, clicks, miaows and snorts. If danger or a threat is perceived they will shriek shrilly and distinctively to warn the other lemurs in the troop to flee and escape.

The two lemurs faced one another and touched noses, but the calm interlude soon ended with the approach of another male who had followed the pair, marking their spots with his own scent. The first male turned on the newcomer and gave a shrill bark, lowered his muzzle and stared hard at the challenger. The new male, who had bald patches on his thighs, squatted down, making mewling sounds in his throat. He rubbed the glands on the insides of his wrists and marked the branch on which he was sitting and then began to move about, always keeping more than a jump away from the other lemur. The dark-eyeringed male persistently marked the spots that had been marked by the bald-thighed male. From time to time the animals paused to stand on their hind legs. They would thrust their tails forwards between their legs so that they could draw their wrist glands deeply through the rich fur; then they would wave their smells at each other with their tails. The stink fight went on for a little over ten minutes before the challenger showed signs of giving up and leaving this strange field of glory. The winner strutted triumphantly and groomed himself as the challenger moved away.

The dominant male turned to the female, who was being groomed by a juvenile female. He cuffed the youngster aside but was immediately assailed by another female who screamed and buffeted him. It took several more minutes and much marking before he approached the receptive female again. This time, after he had chased away another interested male, she 'presented' herself to him, turning so that he could copulate with her from behind. Almost at once, a female leaped on him and began to cuff him away, and he had to chase her before returning to his mate.

Several times the couple were disturbed by attacks of this kind and had to separate while the male routed the interlopers. At last they managed to mate, undisturbed, for about three minutes. They coupled twice more, resting in between and grooming each other.

The troop made clicking and gentle moaning noises before they began to move off in search of more food. They dropped to the ground and walked, with many pauses, to sniff the undergrowth for food. The small clearing was bright with sunshine, and the lemurs, tails waving gracefully above their backs, clicked and grunted to each other as they moved in a fairly compact group to the side of the glade bordering the river.

Once they had crossed the open grassy space, the ring-tails climbed quickly into the trees and searched for hollows and broad leaves where the dew had collected. They refreshed themselves by drinking the water before zig-zagging back across the glade to a **clump** of silikata sifaka. Here they settled to eat the berries on the trees. Pulling a fruiting branch near, the lemurs delicately picked the berries with their mouths.

One of the females had climbed higher than the others. She was near the top of an acacia tree that lifted its head above the surrounding trees. She had found a kily pod on the ground and carried it up into the tree. She cracked a pod open with her molars and scraped the seeds into her mouth with her teeth. When she worked her way down the pod towards its base, she paused to sniff it carefully, then turned away her head and yawned widely as she tossed the rotten part of the pod down to the ground.

A movement caught her attention – less than 100m above, and slightly behind, her a great grey harrier hawk sailed on still wings. The lemur made a mewling noise and leaped down through the branches towards the rest of the troop. They mewed too and then

Opposite page:
Found only in the arid forests of southern Madagascar where the sparse undergrowth enables it to walk on the ground, the ring-tailed lemur is mainly herbivorous in its diet, eating a mixture of fruits, flowers and leaves. Its favourite foods are the kily pods and flowers of the tamarind tree and nearly half of its diet comes from this source. Animal food is rarely taken although this species of lemur is probably the most dangerous of all lemurs as well as one of the largest. It is the size of a small cat and not unlike one in appearance and the miaow noises that it makes. The head of the ring-tailed lemur is very conspicuous with its distinctive black markings and enormous golden eyes. Facial characteristics vary quite considerably between one ring-tailed lemur and another, the shape and depth of the eye rings, nose line and ear tip markings differing between individual members of the same troop. The muzzle is dog-like and the large canine teeth are shown in displays of aggression to other lemurs.

streamed into a patch of dense undergrowth. The tree cover on this side of the glade was thin, and the hawk flew lower to investigate the troop. As the bird made a low pass the lemurs screamed at it. The noise was intense as the animals drew back the corners of their mouths and sent volleys of shrieks towards the harrier. The bird lifted up and came in to land on the acacia tree recently left so hurriedly by the female lemur. As the bird landed, the shrieks subsided into grunts and rapid clicks.

Soon, the ring-tails moved on to feed among the roots of a huge kily tree, taking little apparent notice of the perching hawk. As noon approached the troop swarmed up into the trees and sought a warm siesta spot. They found three suitable trees less than 200m from those they had slept in the previous night. As they arrived they marked the trees, the females rubbing their anal glands against the branches and the males marking with their rumps and the palms of their hands.

For an hour there was little movement. Then a few of the troop fidgeted about to find more comfortable quarters, and one of the females wandered over to the immature male to groom him. The siesta was the most relaxed part of the day for the lemurs. They sat sunning themselves, rarely **quarrelling**, until late afternoon.

About four hours after the beginning of the siesta, an immature female noticed a colony of beetles in the bark of a nearby branch. She examined them carefully, nosing them but not harming them. She made the same yawn, a rejection gesture, as the older female had made when she found the rotten part of the kily pod. She also made chewing motions with her jaws.

The lemurs stirred themselves and, here and there, brief quarrels broke out. Their clicks and grunting increased. Most of the troop defecated before they moved on to look for another feeding site for the evening. Mewing to each other, they left the three trees, some of the lemurs chasing their fellows playfully.

This acrobatic ring-tailed lemur is agile and graceful as it leaps and swings between the high branches of the trees in the dense forests of Madagascar. Unlike the other species of lemurs, the ring-tailed lemur is not solely arboreal and is equally at home on the ground. However, much of its life is spent wrapped around the uppermost branches of tamarind trees.

Once more, the silicata sifaka tree provided them with leaves and fruit for the main part of their meal. The troop covered no more than about 170m between siesta time and dusk. They left their evening feeding site to seek some tall trees for sleeping. As they left the stand of silicata sifaca trees, a small troop of Verreaux's sifaka lemurs appeared 30m away. The two troops paused to stare at one another but, after some mewing and clicking noises, they moved off.

Two enormous kily trees offered the security the ring-tails sought for the night. The females and one of the juveniles crowded together in the fork of the larger tree. There, they groomed each other as the rest of the troop leaped about the branches of the two trees looking for comfortable resting places. They quarelled and occasionally cuffed at their companions in their search, but gradually the scene became calmer and the lemurs settled to nibbling kily pods carried up from the ground or taken from the twigs they still clung to. One of the males began a call that sounded at first like a mew, but halfway through it he threw back his head until his throat was nearly upright. The call 'yodelled' upwards into a musical howl and carried about 1,000m through the dark forest. Other members of the troop took up the howl from time to time, and every hour or so throughout the night the lemurs howled, heads thrown back and mouths pursed for maximum resonance.

The night was not a completely restful time. Occasionally, the lemurs roused themselves to search for a morsel to eat before settling down again to rest. Some of them curled up in the bole of a tree; others crouched, chin on knees, holding on to a vertical branch. Most slept with their weight on their haunches, their bodies touching. In this companionable security the ring-tailed lemurs closed their amber eyes to sleep and prepare for the next day's search for food, its play and its quarrels. Below them, the forest was dark and many of its inhabitants active. Above, the full moon filled the tree-tops with a cool light which caught the bright ringed tails.

With its long striped tail held aloft above its back like a flag, this ring-tailed lemur is marching along the ground. When walking, the lemur always holds its tail erect in a S-shape. Like most prosimians, the lemur can hold firmly on to whatever it grasps owing to the developed flat nails on its toes. Many lemurs have hands and feet with opposable great toes and thumbs which are important in grasping supports when swinging through the branches of trees.

The lemurs of Madagascar

Millions of years ago, Madagascar split away from the continent of Africa and slowly moved eastwards. The animals that lived on the island were cut off the from the evolving and invading competition suffered by their fellows on the mainland and continued to live undisturbed for 50 million years. In fact, the island became a kind of Noah's ark of species that had long become extinct on the African continent. Among these animals, three families of lemurs (*Lemuridae*, *Indriidae* and *Daubentoniidae*) thrived there although they had died out elsewhere in the world.

The habitat that sheltered the lemurs for so long, and which enabled men to observe their ways, is now threatened. The human population of the island has increased rapidly in the last century and man has hunted several species close to extinction. In recent years, laws have been made that ban hunting of some species and limit the capture of others; but the growth in numbers of urbanised islanders, and their access to firearms, means that the lemur is still hunted. Many of the tribes of Madagascar have taboos about the eating and hunting of lemurs which, in the past, have protected their numbers, but these taboos have become less effective with the growth of the urban population. However, the greatest threat to all lemur species is the commercial use of the great hardwood forests where many of the animals are found. Already large parts of their habitat have been destroyed and troops of lemurs have died out. A further threat looms as the search for oil and titanium advances. The opencast titanium mining operation will probably turn great stretches of lemur country into desert in which they have no hope of survival.

Species of lemur

The **mouse lemur** (*Microcebus murinus*) has a grey coat with a white belly. The mouse lemur is one of the smallest primates, measuring only 125-150mm, from head to rump, and weighing between 45 and 85g. A similar animal of another race, *M. coquereli* has darker grey fur with a liberal sprinkling of reddish hairs. It is larger — 250mm from head to rump on average, and with a tail of about 280mm. Both races are able to store fat at the roots of their tails, using it as food which they can absorb while in a state of torpor during the hot season. *M. murinus* is partly insectivorous, and both races eat leaves, honey and fruit.

Dwarf lemurs comprise three species, one of which, *Cheirogaleus trichotis*, is probably extinct. It was last seen about 100 years ago. Both of the other species are declining in numbers and are now rare. The **fat-tailed dwarf lemur** (*C. medius*) and the **greater dwarf lemur** (*C. major*) are of a similar colour and size. From head to rump, they grow to between 125 and 255mm and their tails are of roughly the same length. *C. major* bears two or three young at a time afer a gestation period of about 70 days.

The **fork-marked lemur** or the forked mouse-lemur (*Phanar furcifer*), lives in northern and western Madagascar. Like the mouse and dwarf lemurs, this species is thought to be solitary in its habits. Its head and body length varies from 250-275mm and it has a tail of between 300 and 350mm. It takes its name from the strong markings that run in two stripes from the eyes over the skull to join on the crown of its head and continue along its spine.

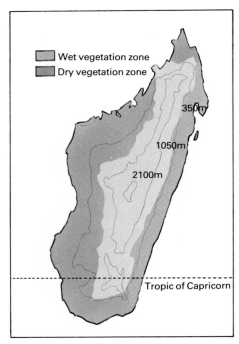

The **weasel lemur** (*Lepilemur mustelinus*) is sometimes called the sportive lemur, and lives in forests in many parts of Madagascar and on the small islands to the north-west. It is mainly arboreal and feeds at night. From nose to rump it grows to between 310 and 350mm and has a tail of 255 to 305mm. These lemurs raise their young in nests made of leaves in hollow trees.

The **broad-nosed gentle lemur** (*Hapalemur simus*) is 280 to 458mm from nose to rump, with a tail of similar length. Its muzzle is short and its teeth (with the exception of the molars) are serrated at the edges for eating the bamboo that forms a big part of its diet.

Lemurs of the genus *Lemur* live in groups of five to 24 in number. They are all tree-living animals, with the exception of the **ring-tailed lemur** (*Lemur catta*): see *At a glance*, panel. Apart from the ring-tailed lemur and *L. fulvus*, all lemurs are rare. The **black lemur** (*L. macaco*) has a distinct ruff. The female of the species differs from the black-furred male in having a brown coat. The **variegated lemur** (*L. variegatus*), unlike other members of the genus, makes a nest for the night's rest. The **mongoose lemur** (*L. mongoz*) has a greyish fur with reddish hairs about the outer parts of its face and a 'cap' of darker hair. These species, with *L. rubiventer*, all eat leaves, fruit and insects.

The **woolly lemur** (*Avahi laniger*) takes its name from the woolliness of its coat. The woolly lemur's face has a covering of short hairs. The animal grows to between 300 and 450mm from nose to rump, and its tail is 325-400mm long. It feeds by night on fruit and leaves and spends most of its time in the trees. When it descends to the ground, it moves in an upright position with its arms held high above its head. Woolly lemurs are often solitary but sometimes congregate in groups of two or three.

Sifaka is a name borne by two species, *Propithecus verreauxi* and *P. diadema*. *P. verreauxi* is found abundantly in the island's deciduous and evergreen forests but its cousin *P. diadema*, is scarce. Sifakas move about in groups of six to ten, feeding during the day time. They are large lemurs, growing to between 508 and 1065mm with tails of 432 to 533mm. Their young are born in September.

Indri (*Indri indri*) look quite similar to sifakas but lack long tails, and their ears are more exposed to view. The animal's head and body grows to between 610 and 712mm long, and its abbreviated tail is only 50-64mm long. Indris rarely descend from the trees, moving about in family groups. After a gestation period of 60 days, an indri female bears one young. She carries it on her back for some months.

The **aye-aye** (*Daubentonia madagascariensis*) lives in eastern Madagascar, where it feeds on insects mainly. Some authorities are undecided whether to include it among the lemurs. Its chisel-shaped incisors which grow continuously, like a rodent's, are used to strip the bark from branches and expose the insect larvae beneath.

Gliding or **flying lemurs** of south-east Asia are not related to the members of the order *Lemuroidea* but are classified in a separate order, *Dermoptera*. The Madagascar lemurs all fall under the order *Lemuroidea*.

At a glance	
Ring-tailed lemur	
Lemur catta	
Class	Mammalia
Order	Lemuroidea
Family	Lemuridae
Genus	Lemur
Length from nose to rump	45cm
Gestation period	134 days
Number of young	Generally one per season
Size of newborn infant	10cm
Longevity	In the wild is unknown, but they live for about 20 – 25 years in zoos.
Diet	Leaves and fruit.

Some of the different species of lemur are illustrated here, including the arboreal mouse lemur which is one of the smallest of the primates. It may measure only 12cm in length from nose to tail and weigh as little as 55g. Unlike the ring-tailed lemur, it is omnivorous, eating insects as well as berries, flowers, bark and leaves. The fat-tailed dwarf lemur lives in the dry forest regions and actually stores fat in its tail, using these fat reserves to live on during dry periods when food is scarce. The white ruffed lemur is unusual in that it is nocturnal in its habits unlike Verreaux's sifaka lemur which is common throughout the forests.

The Giraffe

The faintly damp smells of dew-refreshed grass reached upwards to the dawn as the great sun outlined the horizon. Within moments the dense African night had withdrawn to the far edge of the plain, and long soft shadows unfurled across the rolling grassland to give it an exaggerated perspective. Fingers of sparse woodland stretched like promontories into the sun-yellowed sea of grass that rose and fell in long swells across the plain. A small group of zebra cropped quietly among some thin bushes, while nearby the trees seemed to divide as, with sleepy grace, a tall cow giraffe stepped slowly clear of the cover they afforded.

Her slow-motion walk carried her past the zebra herd towards a clump of shrubs and acacia trees. Their umbrella-shaped tops were still fresh with shoots where antelope had been unable to reach the tender foliage. A scattered group of six more female giraffes with two juvenile males, less than two years old, moved among the fringe of the woodland. The herd was so loosely collected that it looked less like a body of animals than one that had casually found itself in company. They nibbled at the upper branches of the 'whistling thorn' acacia trees. Three of them, standing clear of the others, faced outwards to watch the approaches to the herd on all sides. Their heads swayed gently on their long necks as they surveyed their individual sectors through great, brown, gentle eyes that were fringed with long, thick, curling lashes. These eyes were not only beautiful but the sharpest of any of Africa's big game animals. The zebra stallions continually glanced towards them to make sure that these watchers of the grasslands had not spotted a lurking predator.

The lone cow browsed contentedly among shrubs and trees. Her long, dark tongue curled out to loop round a leafy twig of the tree, just above the 'browse line' where the foliage grew green. Below the line, the trees were stripped bare by shorter animals. She drew the branch between her lips, which were stretched forwards, and closed her mouth on the mass of twigs. Pulling her head back and to one side, she stripped off the leaves and succulent stems with her front teeth. She chewed the wad of foliage with the molars at the back of her mouth; the inside of her cheeks were toughened against the thorns and the stinging insects accidentally included in her diet.

She moved from one tree to another steadily, never spending more than a few minutes at any one of them. *Crematogaster* stinging ants, which had made their homes in galls among the foliage, were shaken by the giraffe's vigorous browsing. They rushed out to attack the disturber of their peace, but the score or so that reached the animal's body could not penetrate the hair and hide of her neck. Those that reached her face discovered that she closed her nostrils against them and swept them into her mouth with her rough tongue. By the time the ants were ready to attack in large numbers, the giraffe had moved to another tree.

The heat of the day grew intense. The air shimmered beneath the craned heads of the sentinel of giraffes which were watching a pack of hunting dogs cross a low ridge about two kilometres away. The dogs paused to look at the zebra herd but when the herd, taking its cue from the watching giraffes, moved into compact order and cantered a couple of hundred metres away, the dogs resumed their patrol and trotted, heads down, from view.

Beside the patch of trees, the cow giraffe gave a soft moo of

Opposite page:
These giraffes are browsing peacefully on the high foliage of some trees against the spectacular backdrop of the mountainous Kenyan landscape. These bizarre-looking creatures move gracefully across the wide savannahs and plains of the African continent in small herds and are instantly recognisable owing to their elongated necks and distinctive markings. The movement and territory of the herds are restricted by the natural barriers of swamps and deep rivers which the giraffes fear to cross. Dense forests also present an obstacle to the wandering giraffe herds.

discomfort. On close examination, she looked rather thicker about the belly than the others, but not very much so. She was carrying her second calf and its time was due. Now labour pains had begun and she quietly chewed more leaves. About a quarter of an hour after the contractions in her abdomen began, a front hoof, soft and swollen so that it would not damage her, reached out from beneath her tail. She strolled about for a few minutes before another soft hoof found its way into the sunlight. The hooves were followed after another few minutes by the calf's head, its soft cartilaginous horns lying flat so as to offer no obstacles to the birth.

The group of female giraffes ambled over to the labouring cow and stood around her looking on quietly. The cow wandered about distractedly while the calf's head moved gently further out and then withdrew a little. This went on for ten minutes. The cow began to move her head up and down as though using her long neck, lever-like, in sympathy with the muscular effort needed to expel the calf. Her calf's black tongue curled out and licked its nose, and its mother braced her legs apart so that her belly was a little nearer the ground. Her neck still moved up and down until, with a jerk, she thrust the calf's shoulders and quarters free and the infant dropped one and a half metres to the ground. The umbilical cord coiled after it until it pulled tight and snapped just before the slippery bundle of the newborn calf hit the floor below its mother.

The calf lay shining wet and still on the hot grass for five minutes before there was a flutter in its chest and it began to stir a little. The little bull calf raised its head from the ground and waved it uncer-

The tallest animal in the world, the giraffe is about five and a half metres high — three times the height of a fully grown adult man. Acacia thorn trees are its main source of food and it browses on the foliage of the higher branches so that the trees often assume an hour-glass shape from over-grazing. Giraffes spend most of their time eating, wandering leisurely from tree to tree and nibbling many different kinds of leaves, twigs, seeds and pods. From their high vantage point their eyes can comb the African landscape in search of predators. Because of its great size the giraffe has few enemies — only lions and leopards.

Opposite page:
The infant giraffe can stand and walk within one hour of birth. Mating and calving take place all the year round and after a gestation period of 420-468 days a single calf is born, already two metres high and weighing 59kg. Giraffe mothers are extremely casual in their treatment of their offspring and although the calf may be suckled on its mother's nourishing milk for up to nine months, it soon starts to eat leaves and twigs as well. It grows fast owing to the high fat content of the mother's milk.

tainly to and fro, touching the ground with its sensitive lips. The cow carefully spread her legs wide, dropped her head and began to lick the calf clean. She stood and walked a little way clear while the placenta broke away and fell from her. While she cleared up this afterbirth material some of the other cows reached forwards to touch the calf with their noses but backed away when his mother returned.

She continued to clean him up until he began to struggle hard, moving his forelegs forwards and to the side. He gathered himself for a great effort, pulled himself on to his long shaking legs and stood, quivering, in the baking air. For some time he moved no further, finally trying a few staggering steps. Within half an hour he managed to walk towards the cow and push his head between her front legs. He reached back to her full udders and took his drink of milk, sucking out the creamy liquid for only a few seconds before the cow backed away from him. Her milk for the first ten days or so would be extremely rich, nearly three times as rich in fats as cow's milk, so the calf quickly felt satisfied.

The cow nuzzled him gently and moved back to the whistling thorn trees to graze while the calf walked unsteadily after her. For the next nine months he would depend for a large part of his nourishment on the giraffe cow, although he would begin to nibble long grass after ten days and browse on low bushes at a few weeks of age. He stood nearly two metres tall, his neck a little shorter in proportion to his body than it would be when he became an adult. Already his hooves were hardening, and in about 15 days his horns would be firmly upright and in place.

The full heat of the afternoon smote the animals of the wide savannah. The sun, broad and white overhead, seemed to press the quivering air to the baked ground. The plain and its herds suffered except for the giraffes. Their long legs and towering necks presented huge areas of surface through which the heat dispersed. One of the females lowered her head until it rested on the lower part of her left hindleg. She slept for ten minutes, breathing gently with her eyes open, before swinging upright again. In the distance two lone bull giraffes ambled slowly towards the females.

Both of the bulls were substantially larger than the cows. When they arrived they briskly herded the juveniles and the newborn giraffe away from the females, refusing to allow them back again. The new mother appeared unconcerned by this treatment of her off-spring. The males began to wander round the small herd of females, sniffing them to find which ones were in heat and ready for mating. Their search quickly centred on two of the cows. The bulls approached them, gently nuzzled their sides, and sniffed the cows' genitals. This behaviour produced in the cows a desire to urinate, and as they did so, the bulls caught a little of the urine on their tongues. They folded back their upper lips and breathed in the scent and flavour of the liquid before ejecting it again in a long stream. This act (called flehmen) helped them to find a cow who was fully ready to mate.

It was soon clear that only one of the cows was attractive to the bulls, and the smaller male sidled up beside the big bull. The two males stood for a few moments side by side, their heads swinging. The smaller swung his head wide and brought it in hard, his horns aimed for his opponent's neck. The older bull easily moved his neck out of the way without moving his straddled legs. As the smaller bull whipped his head back for another try, the big bull slammed in a punishing blow that landed squarely on the other's neck, high

Opposite page:
Giraffes tend to live in loosely structured herds which are rarely stable in their composition and numbers, having a constant turnover in the individual members of the group although they do remain in the same general area for many years. The males live in separate herds in forested zones and visit the females and their young for the purpose of mating. Female herds roam over more open country and scrubland. The male calves may stay with their mothers for up to 12 months before they leave the herd to seek adult male company. Old males are often solitary, preferring to browse on their own at their own pace.

The flatness of the African savannah helps to emphasise the great height of the giraffe trotting across it. Giraffes are found in dry savannah lands and semi-desert south of the Sahara Desert. The giraffe's only living relative is the okapi which inhabits more forested areas. However, they are both even-toed ungulates and are both classified in the Giraffidae family.

The flexibility and movement of the giraffe's long neck can be observed easily in this scene of giraffes eating and drinking at a waterhole. The bizarre long neck and heavy head actually assist movement by acting as a useful counterpoise for the giraffe. The pale buff coats are either boldly spotted or blotched in liver, chestnut or brown and the markings vary enormously within species and individuals. Millions of years ago, the giraffe roamed through Europe, Asia and Africa and many fossilised remains have been found, but they are now restricted to certain areas of Africa only, living in lightly wooded or open country.

The giraffe's great height is a disadvantage when the animal wants to drink and to do this it must straddle its legs widely and awkwardly and bend its long neck downwards. This ungainly position renders the giraffe vulnerable to such predators as lions and leopards which can spring on it, catching it unawares. Although giraffes can withstand long periods without water, they drink frequently when it is freely available. However, they can survive in dry, arid regions, satisfying their liquid needs from the moisture in leaves and foliage.

up. If he had not been moving back, the blow might have ended the fight quickly, but the small male rode the blow lightly and, while the swing followed through, slipped his head down a little and drove in a swipe at the big bull's forelegs, catching him off balance. The big giraffe staggered, nearly fell and, as he tried to sort out his long legs, received a well-timed blow on the neck. That did it. He ambled away, his head bowed in submission. The smaller bull followed him a little way but did not press his advantage. He carried his head high and walked over to the female in heat.

The new-born calf rejoined his mother while the smaller bull mounted and mated his chosen female. The big bull broke into a gallop to chase away the two juveniles, but they were lighter than him and moved away from him at a brisk and graceful gallop, their necks swaying backwards and forwards as they sped across the plain.

The season had been dry and harsh but a waterhole nearby still contained a pool of scummy water which was sought by the animals of the savannah. The herd of giraffes ambled steadily towards it, wary of the danger of predators. The bull slowly bent to drink, spreading his forelegs wide to lower his head to the water and curling it into his mouth with his tongue like a dog. He shared his place at the waterhole with several gnus, Thompson's gazelles and a warthog family. The cow giraffe that he had just mated stood back a little way to gaze across the low vegetation which surrounded the hole, making good cover for stalking animals.

The giraffe who had given birth that morning drank briefly and then drew away from the hole to follow her calf. The sentry cow failed to see the lithe, tawny shape that crouched in the bushes. No movement gave it away. Golden eyes sought out a victim easy to take and nourishing to eat. The lioness had been waiting there for more than three hours, thirsty and her stomach rumbling for food.

The calf wandered towards the low bushes, sniffing the new and interesting smells. The cow had almost caught up with her calf when the lioness saw the tall shape and realised that she must launch her attack at once if she were to kill the young giraffe without interference. With less than 100m to cover, the lioness launched herself

across the broken ground like an express train. The calf was immobile, stiff with fright, but the cow, only a few strides behind, bounded forwards in her slow-motion gait until she was beside the calf. The giraffe reared her enormous forelegs clear of the ground as the lioness leapt for the calf's throat, unable to check her charge even if she had wanted to. The giraffe's forelegs flashed out in a vicious kick of enormous power, catching the lioness in midflight. The sharp hooves struck the predator under the throat and tossed the animal to one side with a dreadful sound of horn smashing through tough muscle.

The lioness was dead before she hit the ground, her head half severed by the rearing kick. The cow turned and trampled the twitching corpse, driving in her hooves with all her impressive weight and strength. Though sweating and trembling, the young calf stood by, watching his mother, unharmed.

Ambling over to the trees, the giraffe group divided. Two of the females wandered off with the small bull, while the rest joined another small herd of females and juveniles in a group of thorn trees a quarter of a kilometre away from the waterhole. As the heat diminished, the giraffes stood resting among the trees. Every so often a lump travelled up their throats and their jaws worked from side to side. The muscles of the oesophagus were flexing to draw up a ball of cud from the rumen, one of the animal's four stomachs, into the giraffe's mouth. After chewing it for about 40 seconds it was swallowed again into another of its stomachs, the omasum.

For much of the night the giraffes would chew their cud, pausing to doze for a few minutes and to nibble a few fresh shoots from the scattered trees. It might be some weeks before they ventured to another waterhole, but they could survive easily without water for two or three months so long as the vegetation was adequate.

As the light failed, the yellow-billed oxpeckers pecked at the ticks irritating the giraffe's hides, perching on their giant hosts with impunity. The night sounds took the place of the day sounds, as, uncomfortably close, a lion roared his impatience into the night.

Opposite page:
Giraffes usually spend the hottest part of the African day standing quietly and chewing their cud. Because they need little sleep, only sleeping deeply for less than one hour per night, they often browse among the treetops after darkness falls. They usually sleep standing up, resting their heads against fellow giraffes or a tall tree. They are at their most vulnerable to attack from predators when lying down, and rising from a seated position can be a lengthy and difficult operation.

These giraffes' distinctively shaped heads are silhouetted against the glowing African sunset. Both sexes have horns, or cartilaginous growths, which lie flat on the head at birth but rise up soon afterwards. They resemble the antlers of a deer when it is 'in velvet', and in the male giraffe they may be up to 25cm long; they are smaller in the female. Sometimes other bony protuberances appear in addition to the two horns, so that some giraffes may appear to have as many as five horns at one time.

Species of giraffe

Giraffes live only in Africa, where their numbers have decreased substantially since the coming of the white man with his sporting rifle to the continent. The natural barriers of swamp, and deep, wide rivers and lakes, and dense forests restrict the natural movements of herds and their colonisation of otherwise suitable habitats. They fear swampy ground, where they can become trapped in mud, and although they have been seen wading in lakes and rivers with their heads rising out of the water, they cannot swim. Fences offer slight obstacles to giraffes, which can jump them, clearing up to two metres without much trouble.

From time to time, since it was introduced in imported cattle, rinderpest, a form of cattle plague, has infected and killed animals across great stretches of the continent, including many giraffes. The distribution of giraffes has altered in historic times as the deserts of north Africa have encroached further south.

Giraffa camelopardalis camelopardalis, sometimes called the Nubian giraffe, was the first to be named systemmatically by 'the father of modern biology', Linnaeus, in 1758. Its legs are white below the hocks and spotted above on the outer sides of the limbs. The body markings are smooth at the edges and widely spaced.

G.c. rothschildi has markings that are, in many animals, interrupted with streaks of the light background colour. The markings are occasionally fan-shaped and may have rough edges, but generally the edges of the dark markings are clear. In the past, this subspecies of giraffe was further divided to include *G.c. cottoni*, but the races are now regarded as one.

G.c. reticulata has dark reddish-brown spots, even darker in old animals. The markings are large and placed close together, leaving a thin network of light colour dividing them. The markings extend well down the legs, even beyond the hocks.

G.c. tippelskirchi has markings of a much-divided design, some leaf-shaped, others tending towards a star shape. The spots may be small and on other animals much larger, even resembling *G.c. reticulata* in size and depth of colour but, unlike the latter, the spots never extend down as low as the hooves.

G.c. thorncrofti has lighter shades of markings on its neck than on the rest of its body. They start as oblongs on its neck, becoming splintered as they descend its body, and extend all the way down its legs. A recently noted variation to the normal is that three *G.c. thorncrofti* have been observed to have three lobes to their canine teeth. *G.c. reticulata* has only two lobes to its canines, and the difference may be a racial characteristic.

G.c. angolensis has spots all over its legs. The markings are clearly defined but have small notches. The clarity of their shapes becomes diffuse towards the top of the neck.

G.c. giraffa is a doubtful race in so far as it may include *G.c angolensis*. The markings are generally very dark and roundish.

G.c. peralta has a hide with a background colour of a reddish ochre among mature adults. The spots continue below the hocks.

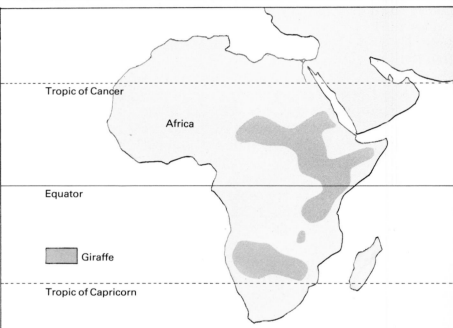

At a glance *Giraffe* *Giraffa camelopardalis*	
Class	Mammalia
Order	Artiodactyla
Family	Giraffidae
Height	Male 4 – 5.88m (the latter measurement was taken from an exceptional bull). Female 3.81 – 5.17m (the latter measurement was taken from an exceptional cow; average heights for bulls and cows are about 4.5m and 4m).
Height at shoulder	Male up to 3.20m Female up to 3m
Tail length	78.7 – 104.1cm with a tassel of up to 1m.
Weight	950 – 1800kg.
Diet	Herbivorous: browses on tree and shrub leaves and shoots.
Maximum speed	56kph
Gestation period	420 – 468 days.
Calf's weight at birth	47 – 70kg
Calf's height at birth	1.7 – 2m
Lifespan	20 years or more in the wild.
Thickness of hide	At the shoulder 15mm. At the neck 8mm.

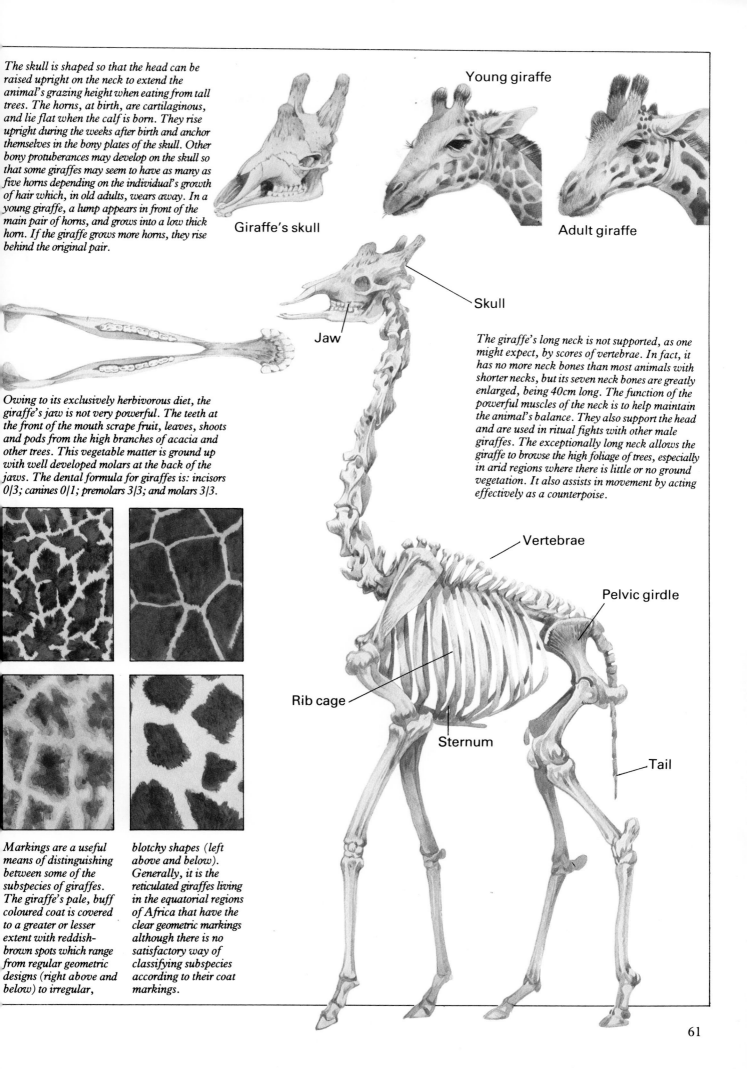

The skull is shaped so that the head can be raised upright on the neck to extend the animal's grazing height when eating from tall trees. The horns, at birth, are cartilaginous, and lie flat when the calf is born. They rise upright during the weeks after birth and anchor themselves in the bony plates of the skull. Other bony protuberances may develop on the skull so that some giraffes may seem to have as many as five horns depending on the individual's growth of hair which, in old adults, wears away. In a young giraffe, a lump appears in front of the main pair of horns, and grows into a low thick horn. If the giraffe grows more horns, they rise behind the original pair.

Giraffe's skull

Young giraffe

Adult giraffe

Owing to its exclusively herbivorous diet, the giraffe's jaw is not very powerful. The teeth at the front of the mouth scrape fruit, leaves, shoots and pods from the high branches of acacia and other trees. This vegetable matter is ground up with well developed molars at the back of the jaws. The dental formula for giraffes is: incisors 0/3; canines 0/1; premolars 3/3; and molars 3/3.

The giraffe's long neck is not supported, as one might expect, by scores of vertebrae. In fact, it has no more neck bones than most animals with shorter necks, but its seven neck bones are greatly enlarged, being 40cm long. The function of the powerful muscles of the neck is to help maintain the animal's balance. They also support the head and are used in ritual fights with other male giraffes. The exceptionally long neck allows the giraffe to browse the high foliage of trees, especially in arid regions where there is little or no ground vegetation. It also assists in movement by acting effectively as a counterpoise.

Skull

Jaw

Vertebrae

Pelvic girdle

Rib cage

Sternum

Tail

Markings are a useful means of distinguishing between some of the subspecies of giraffes. The giraffe's pale, buff coloured coat is covered to a greater or lesser extent with reddish-brown spots which range from regular geometric designs (right above and below) to irregular, blotchy shapes (left above and below). Generally, it is the reticulated giraffes living in the equatorial regions of Africa that have the clear geometric markings although there is no satisfactory way of classifying subspecies according to their coat markings.

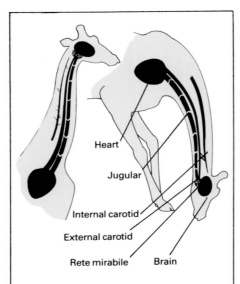

The movements a giraffe makes in raising and lowering its head place great demands on its blood circulation. Its blood pressure when upright is 200mm (equivalent of mercury) and when the animal's head is lowered the pressure falls to 175mm. The danger of flooding the brain's blood vessels is clear. The pressure in this vital area is held fairly constant whether the head is raised three metres above the heart or dropped to about two metres below it. This is accomplished by the *rete mirabile*, a vascular transmission system which lies at the base of the brain. The mechanism works by expanding or contracting arteries to maintain a fairly constant pressure in the cerebral blood vessels. In this way, the neck acts as a collecting vessel for blood, compensating for pressure in the brain.

These male giraffes are indulging in a ritualised form of fighting known as necking. When they fight among themselves, giraffes rarely use the damaging kicks that they sometimes use in defence against attacking predators. Instead, they rub their necks together vigorously and butt each other with their heads. Although necking contests are usually harmless, the blows can be quite painful and, on rare occasions, a giraffe may break its opponent's bones or knock it unconscious with a head blow. The purpose of this ritualised fighting is to establish dominance in a group of male giraffes and is rarely practised by herds of females.

The giraffe can walk, trot or even gallop, reaching speeds of up to 50km/h. Although its usual walking pace is a slow ambling gait, it can gallop fast to escape potential predators. When galloping, the hindfeet overtake the forefeet, passing outside them. However, when the giraffe walks, the two legs on one side of the body move forward in unison, instead of alternating with the other legs. The giraffe's long neck plays an important role in balancing the animal and giving it leverage when moving. It also probably helps to lengthen the giraffe's exceptionally long stride.

Giraffes spend much of their time feeding and can reach leaves up to a height of six metres. Acacia and mimosa trees provide them with their favourite foliage. Their teeth are specially adapted for browsing with wide, lower canine teeth for combing the leaves from trees. They reach around acacia thorns with their prehensile tongues and thick lips which are protected by stiff hairs. They break off the twigs in their mouths, twisting them sharply over the lower front teeth.

The Kangaroo

The plain lay level and mottled with shadows from the low scrub under a sky that quivered with millions of stars. The moon was a smoky-orange as its soft light filtered through the dust-bearing warm currents that had risen from the bush at the end of a burning day. Here and there the shadows clustered darker on the ground, where small clumps of mulga trees formed skimpy groves of cover.

Among these shadows, others moved slowly. There was no pattern in their movements. Sometimes one would move in a series of jerks and then, after a pause, creep slowly to another piece of scrub. As the moon's light grew clearer, and its disc paler, the moving shadows resolved into kangaroos which grazed on the scraps of sun-dried grasses and leaves of the bushes. Near the largest of the mulga groves a small 'mob' of eleven kangaroos grazed the fringe of the patch of herbage that spread from the border of the grove. Here, the shade of the trees in the daytime protected the short sweet grasses growing on a puff of lime-rich soil that formed a mound about two metres above the surrounding ground.

The grass was cropped nearly to its roots, but the kangaroos clipped at it to gain the most essential part of their daily food. Their scissor-shaped lower teeth skimmed the ground to slice off the succulent morsels of grass and herbs as economically as they could. Occasionally they nibbled at the dry spinifex and paused to roll it into wads, packing it together with their tongues in the gap between their incisors and the ridged premolars and molars further back in their mouths. The grinding up of the wads was a slow and nearly continuous process, the kangaroos' jaws sliding against one another to break down the coarse fare. As they chewed, the kangaroos sat up on their long hindlegs, balancing on their powerful tails, while their heads turned steadily from side to side, watching and sniffing the air for danger.

The plain, with its low mounds of limey soil, was not their usual grazing ground – that lay about 70km to the west. However, the dryness of this, and the previous, season and the consequent scorching of the grasslands, drove them to seek for better grazing further afield than their usual range of about 35km. Here, the taste of rain was faintly in the air, and when it fell the plain would abound with life at astonishing speed. The limey puffs that pimpled the plain would be rich with delicious grazing. In the meantime the kangaroos moved about restlessly to try to assuage their hunger.

On the fringe of the mob, a small female joey (a young kangaroo, not yet fully independent of its mother) moved about uncertainly, trying the herbage in her reach while her mother grazed nearby. The joey was small for her age because her mother's milk had not come as regularly nor had been as rich as usual. The dryness of the fodder and its scarcity made it hard for the mother to find enough food to bring her milk on in its customary quantities. Nevertheless, the joey was quite lively, even though she was rather small for an 8-month-old kangaroo.

The mob had been in one spot for some minutes when their eyes were drawn towards the big red who was the most restless member of the clan. He stood a good two metres tall, craning for a better view. His jaws had stopped chewing and his ears swivelled towards the joey who was moving still further west of the mob. The big ears, one slightly tattered from a past tussle with another male, adjusted

Kangaroos live in groups, or 'mobs', which are usually dominated by a large male. In the cool of the mornings and evenings, they graze in the open whereas they lie in the shade of trees or bushes during the heat of the day. At the slightest sign of danger they flee in a series of large bounds and hops which may measure up to nine metres or more. Their powerful hindlegs and slender feet are specially adapted to this curious form of movement. Their stomachs are specialised also for their often dry and inhospitable environment, the bacteria in their stomachs breaking down the cellulose in their diet of plant material and enabling them to digest food more efficiently.

*Opposite page:
These two red kangaroos are resting on all fours in the shade during the hottest period of the day. Placing their weight alternatively on their hindlegs and then their forelegs, they move along in this manner while they feed. Unlike the grey kangaroos which inhabit the wooded country of the eastern states, the red kangaroos graze in their thousands on the arid, open plains and scrublands of the interior, and the northern savannah lands.
They can efficiently utilise the sparse vegetation of this arid landscape and can even survive in times of drought when the network of streams drains away to a few scattered waterholes. The kangaroo can conserve water and shelter from the great heat and thus is perfectly adapted to life in the desert-like regions of the Australian interior.*

their angle to receive any sound from the slight movement that had caught the old kangaroo's sharp eyes.

A shadow had changed shape slightly. It was as if a patch of shrubby growth had slowly grown a longer branch, a smooth branch without any twigs, one that curved sinuously out towards the joey who was scuffling in the dust to look for juicy tubers.

The big red could hear nothing as the Queensland python flowed like a dark stream towards the small animal, but he recognised the danger. With a great spring he took off towards the young kangaroo, landing three metres from his starting point and slamming his tough feet into the ground as he landed. The earth vibrated with a loud thud as he took off on his next leap, and the youngster started upright and began to bounce furiously towards his mother. The python felt the vibrations of the first booming thud and whipped back, shrinking from the big hard feet of the old boomer. The whole mob took off, feet slamming into the baked earth. For more than three kilometres they ran in a straight line until they reached a long wire fence which crossed their course at an angle and they turned to follow it. They swept along its perimeter in graceful arcs, their forefeet held loosely forwards and their tails rising and falling like balance arms, countering the enormous leverage of their hindlegs. The pace of the agile females at the beginning of the run had been faster than 40km per hour, but this had soon dropped to little more than 25km per hour. The group paused several times before halting by a slight hollow in the ground where the wire ran a few centimetres higher than elsewhere above the dusty grass. One of the females squirmed underneath the wire and into the pasture on the other side, and the others followed her.

Sweating slightly and panting from the exercise, the group settled to grazing the edge of a dry river bed, always moving carefully and with their heads raised to scan the plain for danger.

The whole episode in the mother kangaroo's life had occupied only a little more than two hours. She licked the fur through which the newborn kangaroo had climbed to his new home.

The joey, who was now 'at foot', nuzzled at her mother's pouch. Her soft nose entered the pouch and her lips fastened round the teat she had for so long been attached to. The tiny new inhabitant of the pouch was undisturbed, hanging onto his own, newly possessed teat. The extraordinary thing was that the quality of milk for each young kangaroo was suitable for its own individual age and digestion; the mother was producing two grades of milk at the same time.

The heat of the morning increased. The kangaroos panted to cool themselves and licked their forelegs and chests. As the saliva evaporated they lost a little heat; but now they were resting, they did not sweat. They used only those methods of losing heat that were most economical in retaining the moisture so precious to their survival. Beyond the path of the dried river the ground shimmered in the heat so that the horizon divided into layers of mirage. Through these layers straggled a mob of skinny kangaroos. There were only six of them, with no sign of joeys. Slowly the newcomers approached the others and stopped to rest under the low brush on the river banks.

There was little response from the healthier mob, which watched carefully, the males creeping closer to look the strangers over and then returning to their own group. Several groups had collected in a loose association over the last weeks, all waiting for the weather to break in this scrubby country where the grass would spring fast only after a downpour.

The big old male might have been feeling the heat. A young buck turning over one of the old fellow's leftover tubers suddenly caught a hefty clout across the side of his head. Instead of hopping clear as he usually did in the circumstances, he bailed up against a bush and defied the old red with a belching cough. The big kangaroo seemed surprised at this demonstration but coughed loudly and hopped close. For a few moments they boxed with their forepaws, aiming

Big male kangaroos, or 'boomers' as they are often called, can be formidable fighters although they have few enemies apart from man. They learn to grapple with each other, 'boxing' with their forepaws when they are juveniles in boisterous games. However, as they grow older, fighting becomes exceedingly dangerous as a kangaroo may grip his adversary by the forepaws and, resting on his powerful tail, kick him with both hindfeet simultaneously. The sharp claws can cause a nasty gash to his opponent or may even completely disembowell him.

lows at the head. It was unusual for them to fight for so long, but the old kangaroo put a brisk end to the skirmish when he rose up on the base of his tail and drove both hindfeet forwards in a devastating kick to his opponent's stomach. The young kangaroo fell sideways through the bush and scrambled clear, blood oozing through his thick fur where the claws had scored his hide. He was fortunate to have got away so lightly; such a blow was quite capable of disembowelling him, although it was rarely struck at full power.

As the sun reached its zenith, the kangaroos' scratching became almost a frenzy of activity. Many of them drew their muscular tails forwards between their legs and hunched their bodies to present as small a surface as possible to the sun's light. One of the females crept under a projecting tangle of roots and mud which had dried to form a shallow cave in the river bank. All the kangaroos licked and panted while they endured the blistering heat. The air was dense with the hum of insects.

In size and build, both red and grey kangaroos are very similar. The reds are larger and derive their name from their dusty, brick colour, although the females are lighter and sometimes may be a blueish-grey. Being exclusively herbivorous, they live in open grassland and compete with sheep for grazing land, although they crop the grass more closely with the sharp teeth in their upper and lower jaws. They often live in arid regions and can exist without water for long periods, deriving sufficient moisture from the vegetation they eat. As well as being good jumpers, they are excellent swimmers and can dog-paddle their way across deep creeks and wide rivers.

The rest of the night passed quietly for the kangaroos and, as the stars paled in the light of the dawn, they lay down or crouched to scratch or to chew a wad of fodder. The sun lipped the horizon and lanced its rays across the landscape. The warmth seemed to rise at once out of the hard soil, gently at first, but the kangaroos loafed about with bodies still cold from the night. Their temperature was as much as 2°C below normal, but this gave them a start in resisting the heat to come as the sun continued its journey up the sky.

The light reflected the colours of the tawny red males, whose chests were stained almost claret-red by a powdery secretion from glands at their throats. One of the young males was not yet mature enough to sport this signal of his readiness to mate. The pale marks that ran from his whiskers to his eyes had a blackish stripe through their centre and were fresher-looking than the markings of the older members of his clan. Like all the kangaroos he had clusters of well-rounded ticks clinging to his body. The big red, largest and oldest of the mob, had a bunch of them sucking his blood at the base of his ears.

The irritation of the ticks and other parasites that infested their hides kept the kangaroos busy scratching. The joined second and third digits of their hind feet were especially useful for delicate scratching. The short claws combed out the loosest ticks and burrs, but both fore- and hindfeet were brought into action in an attempt to assuage the constant itching.

As the morning wore on the females remained gracefully stretched out on the dusty ground. Their soft blue-grey coats were hard to distinguish from the surrounding colours at a distance of over a hundred metres. The males squatted on their tripod arrangement of hindlegs and tail, more alert to any sign of danger than their females

The young joey remains in its mother's pouch for about eight months after birth. The birth itself is an extraordinary feat, the newborn baby being blind and deaf and weighing only 0.9gm. Born at an early stage of its development with many of its organs incompletely developed, it crawls slowly up through the mother's fur, gripping with its well-developed forelimbs, to the pouch where it continues to grow. Before the birth, the mother cleans her pouch in readiness for the joey's arrival, sitting on the base of her tail with her hindlegs extended in front of her. As the newborn kangaroo makes its tortuous way up into the warm pouch where it attaches itself to a teat, the mother licks away the afterbirth substances.

sheep. The culling figures for kangaroos are increasing as their meat is more widely eaten; there is a safety factor in this as well as the obvious danger of over-cropping. If farmers see the advantage in converting their pastures, or parts of them, to raising kangaroos, the danger of the kangaroo's eventual extinction will decline.

Species of kangaroo

Great grey kangaroos, as their name suggests, are grey, but their colour varies slightly between species. The great grey forester (*Macropus giganteus giganteus*) is silvery-grey to rusty-grey, both male and female being the same shade. They move with their heads held high and their forequarters low. Their tails curve upwards when they are hopping. Males grow to between 160cm and 240cm, and they have tails of 75cm to 100cm in length. There is little difference in size between the sexes.

The western grey kangaroo (*M.g. fuliginosus*) varies from light grey to a chocolate brown. As it hops, its tail whips up and down like a pump handle. Males and females are of the same colour, but the males are distinguished by a strong smell.

There is a grey kangaroo that is confined to Tasmania, the Tasmanian great grey kangaroo (*M.g. tasmaniensis*). There are two other species of grey that are widely recognised by naturalists, the southern great grey (*M.g. major*) and the great grey kangaroo (*M.g. ocydromus*), which lives in an isolated pocket of the south-west of Australia.

Red kangaroos are red to grey in colour. There is usually a blueish tinge to the coats of the females, and they have pale abdomens in contrast to the light brown bellies of the males. The task of distinguishing these animals is made more difficult because these colour characteristics are sometimes found in reverse: the males may have a blueish tinge to their coats, and there are females with reddish fur. The males grow much larger than the females.

There are three generally accepted species of red kangaroos: the red kangaroo (*Megaleia rufus rufus*), the western red (*M.r. dissimulatus*) and the northern red kangaroo (*M.r. pallidus*).

The **walleroo**, or euro, is found in most parts of Australia where there are gullies, rocky outcrops and ranges. The walleroos have long, coarse hair, dark forefeet and hindfeet, and their tails are tipped with dark hair. The colour of their fur varies from red-brown and blue-grey (which looks black from a short distance) to fawn. They are thickset animals, and they move with their forelimbs held close to their sides. They grow to a length of 135-240cm.

There are three main species of walieroos: the New South Wales walleroo (*Macropus robustus robustus*), the deer walleroo (*M.r. cervinus*), a walleroo of the central and southern plains (*M.r. erubescens*) and the antelope walleroo (*M.r. antilopinus*), which inhabits the plains and the broad valleys of northern Queensland, Northern Territory and North Kimberly of Western Australia. The noses of the three main types of kangaroos offer a clear way of distinguishing the reds, the greys and the walleroos from close quarters. The grey kangaroo has the hairiest nose, with only a narrow band of bare black skin around each nostril. The red kangaroo has a middle-sized, bare, black area, rather like a broad, shallow 'V'-shape, above the nostrils, whereas the walleroo has the barest nose, the upper part being hairless and black.

Birth

The pregnant red kangaroo has distinct advantages over most other pregnant mammalian species. Most female mammals have to carry an embryo or several embryos of considerable weight, together with its feeding system (placenta) and protective fluid, while living with all the stresses of life in the wild. The kangaroo, on the other hand, carries a minute embryo with its similarly small life-system for a short length of time. The growth of the young kangaroo inside its mother is so slight that it causes no detectable change in the female's shape or her capacity to seek food and run from trouble. After the birth of the baby, at a stage of development when it is barely recognisable as a kangaroo, its weight in its mother's pouch is negligible.

A few days after giving birth to her young, a female kangaroo mates again. The fertilised egg remains in her uterus, lying dormant as a kind of reserve in case the young in her pouch should fail, through weakness or accident, to mature. If this should happen, the dormant embryo begins to develop in the usual way. If all is well with the suckling young kangaroo in the pouch, the process of suckling will hold up the development of the dormant egg until the pouch is ready to receive the new offspring about a day after the firstborn leaves it. The kangaroo is a highly effective year-round breeder. This sequence may be interrupted, sometimes for long periods. In times of severe drought, kangaroos will stop breeding altogether, but a dormant egg carried by a female may develop even after many months, so if a female becomes isolated from males in her group she may still bear a young kangaroo. One female in captivity astonished her keepers by producing a baby after being isolated from males for over 300 days, ten times the usual gestation period.

The first two years

One young kangaroo may develop at a different rate to another of the same species, depending on the general health of the mother, and the quality and amount of food available to mother and offspring. In unusually dry conditions, the female kangaroo will have to eat the dry, stringy grass that survives and consequently will not produce the rich milk essential for her joey's growth.

The newborn kangaroo has a large mouth and nostrils but its ears and eyes are still covered by skin. Claws start to grow on the well-developed forelimbs and the hindfeet at 12 days and sex soon becomes detectable. Between 31 and 47 days, the papillae appear, ready to grow whiskers, and after 70 days the youngster can detach itself from the teat for the first time. Soon afterwards it starts to squeak and whiskers and body hair begin to grow. At the age of 144 days the eyes open and six days later the joey pushes its head out of the mother's pouch to observe the outside world for the first time. However, it does not leave the pouch for another 40 days and then only for short periods as it learns to eat solid food. It returns to the pouch to rest and sleep.

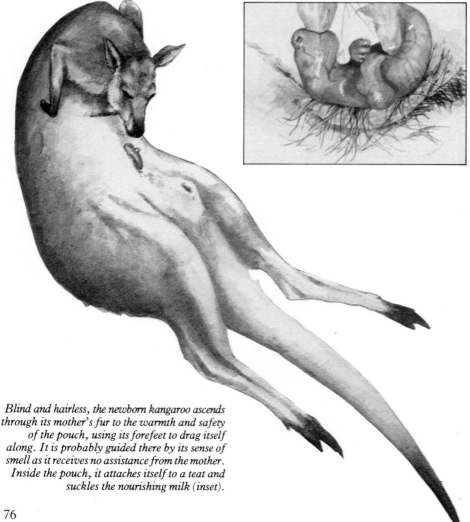

Blind and hairless, the newborn kangaroo ascends through its mother's fur to the warmth and safety of the pouch, using its forefeet to drag itself along. It is probably guided there by its sense of smell as it receives no assistance from the mother. Inside the pouch, it attaches itself to a teat and suckles the nourishing milk (inset).

The joey gets into the mother's pouch head first, turning a somersault once inside so that its head, tail and hindlegs protrude through the opening. In order to do this, the joey uses its hindlegs for leverage.

The body and its fuel

The kangaroo has adapted to its often challenging environment in a number of effective ways. The animal is a herbivore: it eats grasses, shrubs, herbs and occasionally the leaves of trees, and where this fodder is green and succulent enough, the animal needs to drink little water. When the vegetation is burnt by the sun and dried by the wind, it may need to drink about five per cent of its body weight of water.

The slender limbs and the slow metabolism of the kangaroo help it to resist the high temperatures of the Australian desert. It also loses body heat in the night so that it can begin moving on a hot day while its body is still two or three degrees below normal. During the day, the kangaroo assumes a hunched position so that it presents as small a surface to the sun as possible. It draws its tail between its long legs, and the complex arrangement of blood vessels along its body help to dissipate heat. The animal's fur, too, is extremely dense and helps insulation.

Digestion

Kangaroos have a highly effective digestion. Like all mammals, the kangaroo cannot make the enzymes that are needed to break down the cellulose that is the main constituent of grasses. This essential part of the digestive process is performed by bacteria and protozoa in the animal's foregut. Here, the food ferments before passing to the next stage. As the protein breaks down in the body, urea is formed. This is recycled (passed back into the fore-stomach) in the animal's saliva, and micro-organisms in the fore-stomach change it into nourishing protein, saving its goodness from being lost immediately through the urine. The recycling process enables the kangaroo to conserve the water that may mean the difference between life and death in times when drought conditions prevail.

Teeth

The kangaroo is a selective grazer, nibbling off shoots and seeking short green grasses with its incisors. Between the incisors and the premolars there is a gap where the kangaroo rolls the greenstuff into a suitable wad for its molars. The space allows the animal's tongue to operate easily without too much risk of nipping it. The kangaroo lives with the same set of teeth for the whole of its life, with the exception of one premolar which is replaced. The fourth upper and lower molars do not grow until the kangaroo is about four and a half years old.

At a glance
Red kangaroo
Megaleia rufus

Class	Mammalia
Order	Marsupialia
Family	Macropodidae
Subfamily	Megaleia
Length of male	219cm
Length of female	179cm
Length of tail	Male 110cm
	Female 77cm
Weight	Male 77kg
	Female 36.5kg
Gestation period	33 days
Young	Females bear single young but occasionally produce twins.
Weight at birth	About 1.5g
Lifespan	Uncertain in the wild, but probably 8 - 10 years.

The kangaroo's hindlegs are specially adapted to its hopping gait. The greatly elongated foot, powered by the large muscles of the animal's haunches, gives a high degree of leverage. The whole of the foot is placed on the ground in movement so that the animal is said to move in a plantigrade manner. The long fourth digit bears a strong claw, which is used in fighting, and the smaller second and third digits are joined by connective tissue. These small claws are used for grooming the fur. The hindlimb itself is extremely large and powerful for jumping distances. The pelvic girdle and the tibia are strong and elongated. The forequarters are lightly built.

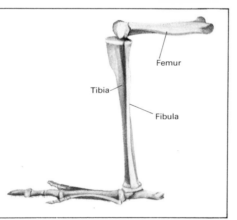

Kangaroos have two gaits: creeping on all fours and hopping. Hopping is an efficient means of locomotion for the animal and it can thus travel at speeds of up to 40km/h to escape its enemies. This sequence shows the different stages in this leaping motion. Using its tail as a foot when the hopping gait begins, it is gradually raised as the body is lifted off the ground to balance the forepart of the body and act as a lever. A spring-like extension of the hindlegs is followed immediately by a squatting contraction in readiness for the next leap. As the animal increases its speed the hops become longer and the tail and body are further extended even though the number of hops per minute remains nearly constant throughout.

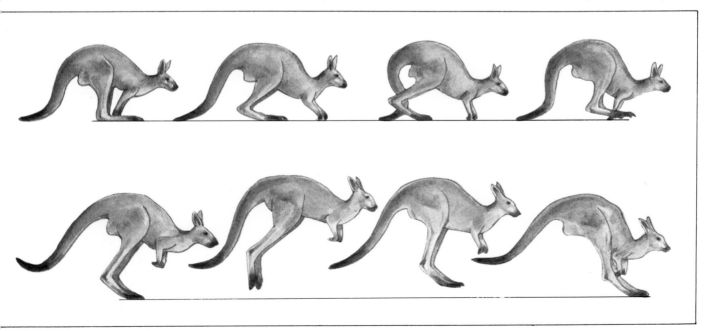

The African Hunting Dog

The huge sun lapped over the horizon to sharpen the outlines of the wide plain. Little movement could be seen at first but a martial eagle quartered the sky, vigilant for any incautious movement on the dusty grass below his outstretched wings. His gaze flickered back on its track, drawn by a sign of life on the edge of a patch of shadow. A small, dark head poked out into the sunlight and, nearby, a tawny, black-and-whitish patch of the ground broke into movement as a dog materialised, raising its head to the sky to sniff the morning air.

A small pup was suddenly propelled from the den by the pressure of other energetic youngsters behind him. They tumbled from the earth, rolling on one another and biting at the tails nearest to them. Their dark coats were blotched with yellowish brown which camouflaged them when they were still. The boisterous play stopped as one of the pups squealed sharply. He had a rather tattered right ear and the other one was firmly clamped between the teeth of a larger reddish-coloured pup. His ear hurt badly and he could not shake his tormentor free. For a short time the other pups watched the uneven struggle then, at once, all joined in to mob the victim. They nipped him wherever there was an available patch of skin until their scrambling shook him clear of the ruck and he bolted between the legs of a large, reddish bitch who was just emerging from the den. With his mother in the way, he could lick his wound in safety.

The pups turned their attention to their mother. They ran up to her whining for attention, pushing their noses against her belly and searching for the teats that had given them milk, but she trotted away from them and lay down. She had not suckled since they were about six weeks old, and that was a month ago. The bolder pups pushed their noses against her mouth but she good naturedly cuffed them away from her.

Other dogs were still lying in their holes or on the veldt itself. The reddish-coated bitch stood up and began to pace around the

Opposite page:
African hunting dogs are naturally gregarious, living and hunting in packs of between 12 and 50 dogs. There is a great deal of social interplay between the members of a pack and they are very friendly in their relations with one another, especially before a hunt when the dogs playfully chase each other and 'twitter' greetings. They hunt at either dawn or dusk across the great savannah lands of central and southern Africa in search of large prey. The hunt and the chase may stretch over many kilometres of territory.

Related to the wild bushdog of South America and the dhole of Asia, the African hunting dog, or Cape hunting dog, as it is often called, is different from most other dogs in having retained five digits on its forefeet. Almost as tall as a wolf but less heavily built, it weighs 27-45kg and has a short, sleek coat with irregular markings of white, brown and yellow blotches. The only constants are the dark muzzle and white tail.

These two African hunting dogs are watching a string of zebra pass on the open plains. The dogs prey on zebra and other large, hoofed animals of the African plains, including wildebeest, buffalo and gazelles. Their persistence, ferocity and stamina enable them to attack and kill prey animals which are much larger and stronger than themselves. The pack usually works as a team, running their prey down, especially weak young or aging animals, until it tires and then closing in for the kill. They rarely scavenge off the kills of other animals, such as leopards and lions.

den. She whined softly at the entrance of another den, and after a short time a large male emerged. Lowering his body slowly between his forepaws while keeping his hindlegs straight, he stretched luxuriously. The bitch nuzzled his head, making a curious high-pitched twittering sound. He soon seemed to pick up her excitement and they trotted to the other wakening dogs, pressing their muzzles to the sides of their companions' faces and licking one another's lips.

The pack was soon in a frenzy of excitement, licking and twittering and rushing to greet each other. The dust flew around them. The pack had a complement of 16 adults – seven females and nine males. The litter of eight pups were all that had survived from the ten born to the reddish bitch, two having died of distemper. The father, a powerful dog with a long scar running from his torn right ear to his muzzle, moved away from the pack, sniffing at the bushes and at the breeze that lightly touched the tussocks of grass. He broke into a steady trot and before long the still twittering pack followed him, stringing out in a line.

A bitch stood watching the file of hunters trot away from the dens. Soon the pups would be ready to follow the hunt, but for now she stayed to guard the litter from other predators of the grasslands, such as hyenas and lions. Once the pups were strong enough, and in another week or two they would be, the pack would abandon their dens for the nomadic life that they usually led except when a litter was being raised.

The file of African hunting dogs trotted in the wake of the scar-faced male, their eyes scouring the plain for prey, their great ears cocked for a sound that might betray the presence of game. For more than two hours they trotted, pausing only when the leader stopped to sniff the ground and mark their passage with a few drops of urine. Game there was in plenty, but most moved away while the dogs were too far off to begin a chase.

A sharp squeal pricked the air and saucer-shaped ears twitched to catch the sound. The leader's head stretched towards the cry, and his neck craned forward level with the ground. He stalked slowly forwards a few paces towards a clump of bushes in a hollow. The shaking undergrowth betrayed movements within the patch. At once the pack knew what was there and broke into a run.

Sounds in the thicket ceased for a moment, and then a large female warthog broke clear. She ran a few metres and paused to let her three young scurry after her. With her tail raised like a flag above her bristling back, she, together with her brood, ran helter skelter for a low bank 200m distant.

It was a close call for the young warthogs. They fell inside the shelter of their den, a hole excavated from the side of the bank. The sow pivoted in a swirl of dust and flying pebbles, skidding backside first after her offspring into the entrance of the den, blocking it. The yelping dogs fanned out round the hole which was now guarded by the formidable head of the warthog. The only way in to the succulent young lay past those curving tusks and grinding jaws. With her body hidden down the hole, there was no way to outflank the animal. The pack was aware of its failure but made an attempt to lure her from her safe position. Their yelping, snapping attacks merely made her lash right and left at them with her tusks. Drawing off a little way did no good either. The warthog wisely stayed in her hole.

The sun rose higher in the African sky and the day grew hotter. Once more the dogs patrolled their range in file. Over to the west the ground fell away in a series of shallow waves down to a waterhole. The summer sun had evaporated most of the water but a muddy patch in the middle still drew thirsty animals. The dogs approached in a wide curve which took them through a patch of scattered and stunted trees. The pack was quiet now, the dogs' tongues hanging out to sweat away some of their heat. About 400m ahead of them, a Thomson's gazelle broke cover and bolted towards the herd grazing warily half a kilometre further off. Her springing gait made her seem to skim the ground, only descending to flick the earth with her elegant feet and take another bound forwards.

These African hunting dog puppies are being suckled by their standing mother. She is helped in rearing and feeding the young pups by all the other members of the pack, which bring back food from hunts and regurgitate it for the young to eat communally. In the pack, the nursing female dogs occupy a special dominant position over other members of the group, living in dens or burrows which afford some protection from the fierce heat of the sun and the savannah winds.

These African hunting dogs are temporarily joined together in the characteristic mating tie position. They usually breed once a year and although there is no fixed season for mating, the cubs are usually born between the months of April and June or August and November after a gestation period of 70-71 days. Other wild and domestic dogs have a gestation period of only about 63 days. The male and other members of the pack help the mother in raising and feeding the puppies which although helpless at birth, soon grow to become playful and boisterous.

The pack did not waste effort chasing her. They would not attempt to run her down unless they could begin the chase at a distance of less than 300m. But why was the gazelle separate from the herd? She was moving well, with no sign of weakness.

Curious now, heads stretched purposefully level with the ground, the pack trotted quickly towards the spot where the gazelle had been hidden. There, before them, a small sickly fawn trembled in the tussocky grass. With excited yelps the dogs closed in with a rush. The fawn died in a few short moments and was eaten in four minutes. Two ten-month-old bitches had a tug of war with a piece of its hide until it disintegrated. Soon, only a few bones and some skin were left for the scavengers.

The sun was high in the sky and the plains sang with its heat. Prone on the earth, the dogs panted, their eyes closed against the flies that seemed to be the only kind of life able to assert itself in that hard bright light. Not until the late afternoon did the pack show signs of stirring. As the cooler air refreshed them they became aware of the need for food. The fawn was only an appetiser.

One of the bitches had come into season a few days before this hunt and she was ready to mate. A rangy male with only half a tail had spent those days as close to the yellowish bitch as he could. When another dog approached, he always interposed his own body between hers and the interloper's, and if the hint was not taken he attacked until the challenger moved away. This afternoon as the sun slid down towards the horizon and as the heat of the day became more bearable, the bitch moved away a few paces to urinate. The male followed her to urinate on the same spot as a warning to other males that she was his mate. He was so eager to 'mark' her urine with his own that, standing beside her on his forefeet and raising both his hind legs in the air, he urinated simultaneously with her.

Circling round her, the dog drew back the corners of his mouth in a long grin and pawed the dust. He made a brief sally towards the bitch, punching his forepaws at her breast. He wheeled away and ran off in a strange manner. The stump of his tail was tucked between his legs and his back was strongly arched. From a distance of a few metres he turned to look at the bitch with a huge grin on his face and he wagged his stump of tail so hard that ripples of effort were visible much of the way up his backbone. He rushed up to the bitch again and mounted her. Their coupling lasted little more than a minute and finished with a display of nibbling and nuzzling.

The other dogs, aroused from their afternoon siesta, greeted each other in the same ritual they had used in the morning, twittering and licking one another's mouths. The frenzy of arousing the pack to its duties in the hunt was soon completed, and the scar-faced dog led them off in single file again.

The trotting pack had covered only about three kilometres when the leader sighted a small herd of brindled gnu in a bowl-shaped depression. The dogs slipped rapidly over the skyline and circled round to approach the herd obliquely. All at once the gnu were startled into movement. The ground shook as the gnu bolted across the valley. The pack, with excited yelps, tore after them, their white-tipped tails flashing their positions through the swirling dust.

In the pack, individual needs are subordinate to group needs and certain codes of behaviour are strictly observed although there is no rigid rank order. A leading dog usually emerges from the ranks at the apex of the loose hierarchy to lead the chase of prey animals and keep order among individual pack members although they are generally non-aggressive towards each other. Feeding is communal and this strange ritual probably derives from the habit of begging for food as puppies. Partly digested food is regurgitated by the pack members for each other and passes from dog to dog, as each begs another by licking its lips and nudging its muzzle.

After two kilometres it began to look as though the herd might break away from the dogs despite the furious 50km per hour pursuit. The dogs had spotted the weakness in one of the immature gnu and, as it ran out to one side from the rear of the galloping herd, they concentrated on running it down. The pack had strung out by now, the scar-faced dog having fallen back and three of the fastest dogs out on their own. In his wake, about 200m behind, came the main bunch of the pack straining to catch up. Right at the back was an old dog who was making hard going of it.

The first of the fast dogs snapped at the gnu's haunches, making ragged wounds from which the blood flowed, and with it the remaining strength of the exhausted animal ebbed away. Another of the fast dogs ran out to the right, and the leading attacker turned the gnu towards this flanking attacker. She leapt at the gnu's head and, avoiding the sweep of the sharp horns, caught him by the nose. The weary, half-suffocated gnu slowed to a staggering halt, still trying desperately to shake his head free. The other dogs seized him by the muscles inside his thighs, hamstringing him.

As he sank to the ground, the main pack arrived and ripped open his flanks, biting through to the vital organs. By this time he was in deep shock and beyond much pain. The dogs ate with amazing rapidity and were gorged by the time a large pack of hyenas arrived to harry them. After a couple of brief sorties to scare the intruders away, the dogs abandoned their kill to the scavengers.

The light was a smoky red as the sun died in the west. The dogs trotted wearily but replete back to the pups they had left early in the morning. As the pack came into sight of the waiting pups and their guard, they were greeted with a rush of desperately famished pups whining for food. The pups and the guard ran up to the hunters and pressed their muzzles to their mouths as a signal to the dogs to regurgitate some of the meat they had eaten. Without any resentment the males and females regurgitated for the pups and for the guard dog. The twittering of the greetings and pleas for food gradually died away as the animals settled down once more to rest under the blaze of starlight.

These two African hunting dog pups are playing a game of 'tug of war' with a hunk of meat. At an early age, some pups emerge as being more dominant than others but their rough, boisterous behaviour and games are usually good-natured and harmless. They soon learn to run and hunt with the other adult dogs and are trained by the older dogs in the arts of hunting. However, it is some months before they have the stamina, strength and skill to participate in a drawn-out chase and the kill of a very large prey animal.

When a pack of dogs discovers a herd of wildebeest, two of the leading members of the pack (left) will select a single animal and chase it, cutting it off from the rest of the herd. The remaining dogs take up positions in a large circle around the fleeing animal to block its path if it tries to double back or veer off to the side. When the leading dogs finally catch up with the animal as it begins to tire and weaken they bite its hindquarters persistently as the rest of the pack closes in to pull the animal down (below). Sometimes the wildebeest is disembowelled and partially eaten while it is still alive. The victim is usually devoured on the spot and may be completely eaten without 15 minutes by hungry dogs. To maintain the chase, the dogs sometimes work in relays, taking it in turns to chase and rest until the animal weakens.

The wild dogs

There are five species of wild dogs in the world and one, the dingo, that was probably a domesticated breed which has reverted to a wild state in large numbers, although these are now declining. The African hunting dog's reputation for ferocious greed was widespread until recent years when a number of naturalists, not least Hugo van Lawick, studied its habits and gave people a much more favourable picture of the dog's place in the natural food chain. This valuable work came in time to check the wholesale slaughter of the animals. Much the same happened in the history of the Asiatic wild dog and the bushdog. The raccoon dog is now increasing in numbers since the Russians imported the strain from Japan.

The **dingo** (*Canis dingo*) is a yellowish-red dog which is a solitary hunter, although it occasionally hunts in pairs. Many people believe that the species was brought to Australia about 40,000 years ago by the Aborigines and that it may be descended from the pariah dogs of India and Egypt. The dingo became a pest after the introduction of the rabbit to Australia. The dingo's numbers grew rapidly as it discovered this easy new prey in profusion. Domestic dogs have bred with the dingo and now the original type of dingo is comparatively rare. The dingo stands 50-55 cm at the shoulder and weighs 20-24kg.

The **Asiatic wild dog** (*Cuon alpinus*) lives and hunts in packs of 3-30 animals. In India it is known as the *dhole*. It is reddish-brown and has a thick under-fur which provides good protection against wet and cold. The packs hunt bears, wild boar, musk deer, spotted deer, markhor, gaur and banteng, and in one instance have been known to kill a tiger. The Asiatic wild dog measures 76-100cm from nose to rump. It has a tail of 28-48cm and weighs 14-21kg.

The **bushdog** (*Speothos venaticus*) is tawny orange and brown with black, or nearly black quarters. Built rather like a terrier, it is adept at catching small mammals, especially the paca (*Cuniculus paca*), which it hunts mostly at night. From head to rump it is 57.5-75cm long and has a tail of 12.5-15cm. It weighs only 5-7kg. The bushdog hunts in small packs of up to a dozen dogs.

The **raccoon dog** (*Nycterentes procyonides*) is the smallest wild dog. From nose to rump it is 50-55cm, has a tail of 13-18cm and weighs up to 7.5kg. Raccoon dogs live in family groups of 5 or 6 animals in rocky banks near rivers and lakes. Their fur is dark brown with pale grey on the head and neck and dark patches round the eyes. The raccoon dog is generally a solitary hunter with a strong preference for fish, which it usually hunts by night. The females have a 52-79 day gestation period and bear litters of 6-8 pups.

The **African hunting dog** (*Lycaon pictus*). See *At a glance* panel.

The African hunting dog is related to several other species of wild dog although they may differ greatly in their behaviour and physical characteristics. The dhole of the Asian tropical and mountain forests is also a pack dog, living and hunting in groups of up to 40 members. They work in relays as a team to run down deer and wild pigs. Like the Cape hunting dog, they hunt in the early morning before the heat of the day becomes too intense. The Australian dingo, in contrast, hunts and lives alone or in small family groups, attacking sheep, kangaroos and cattle after long, exhausting chases across the outback. The raccoon dog of Asia is also less gregarious.

Dingo

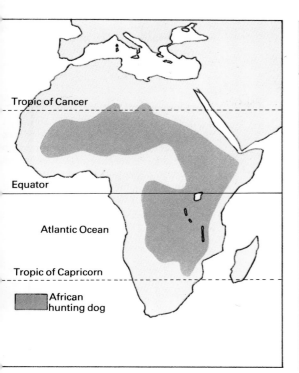

	At a glance *African (or Cape) hunting dog* *Lycaon pictus*
Class	Mammalia
Order	Carnivora
Family	Canidae
Sub-family	Sinocyonae
Length, nose to rump	76 – 108cm
Length of tail	30 – 41cm
Height at shoulder	61 – 75cm
Weight	Male 18 – 28kg
	Female 17 – 24kg
	East African animals of this species are generally a little smaller than this.
Gestation period	69 – 73 days
Number of pups in a litter	6 – 10, although as many as 16 have been reported in exceptional litters.
Coat	The black, tawny and white patches are irregular and vary greatly between individuals, which makes pack behaviour and structure easier to study. The dogs have blackish hair round the throat and a black line running from between their ears down to a point between the eyes. The tips of their tails are nearly always white.

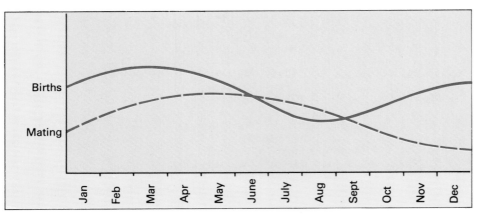

This diagram charts the relationship between the incidence of mating and the number of births. The efficiency of the pack in caring for the young is important in the maintenance of numbers as the mortality rate for adults is fairly high. Thus, about 35 per cent of a pack consists of young.

Like other members of the Canidae family, the African hunting dog is digitigrade — that is, it walks on its toes. This is a characteristic of many animals that have to run quickly over considerable distances in pursuit of their prey. The feet of a hunting dog and a domestic dog are shown here. Whereas the domestic dog still retains a vestige of a fifth toe, that of the African hunting dog has been lost in the evolutionary process.

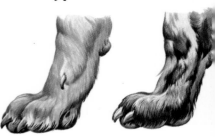

Domestic dog   Hunting dog

Births and offspring

Before the birth of her litter, the bitch prepares a den which will shelter her offspring in the first weeks of their lives. She often selects an old hyena den which she cleans and repairs if any earth has fallen in.

Her young are born with black and white spots and blotches, but over a period of six weeks the characteristic tawny, black and white blotched coats grow. At three weeks of age the pups clamber about, and inquisitive noses can be seen peeping from the entrance to the den. It takes only a shadow from a passing vulture or a wind-tossed clump of dried grass to alarm them and send them scurrying underground again. For several months they are frightened of any unusual movement or threatening sound, responding by ducking back into the den or clustering together near the bitch or another adult in the pack.

About two and a half months after the birth of the pups, they leave the den and travel with the adults of the pack. When the pack makes a kill, the mature animals stand aside while the pups eat their fill first, chasing off any hyenas that might try to steal the prey. The pups often have to struggle hard to stay with the wide-ranging pack, and the weaker ones, despite the occasional help of adults, fall behind or get separated and killed by other predatory animals. Many of the pups die of canine distemper, a virus disease.

The dingo

The dingo, an Australian native dog, is closely related to the Cape hunting dog, and although it has been domesticated to some extent by the Aborigines, it is still found in the wild. It is usually a yellowish-brown in colouring although it may be red or brown or even albino. The dingo presents the farmer with a problem in so far as many sheep and cattle are slain each year by the wild dog. Large bounties are offered for rogue dingoes which kill out of blood lust rather than for food. Because they are listed as noxious animals, bounties are often paid for their scalps.

The dingo (left), one of the purest surviving breeds of dog in the world, hunts mostly at night, either alone or in pairs. Its main prey consists of small marsupials, such as bandicoots (above left). Sometimes, however, it will chase large kangaroos, cattle and sheep, harassing the weaker animals until they slow down and are effectively cut off from the rest of the herd or group. Female dingoes give birth to litters of six to eight puppies (above right) after a gestation period of about 63 days. After being suckled and sheltered in a den for the first two months of life, the pups stay with their parents for the first year. During this period they learn to hunt with the rest of the pack.

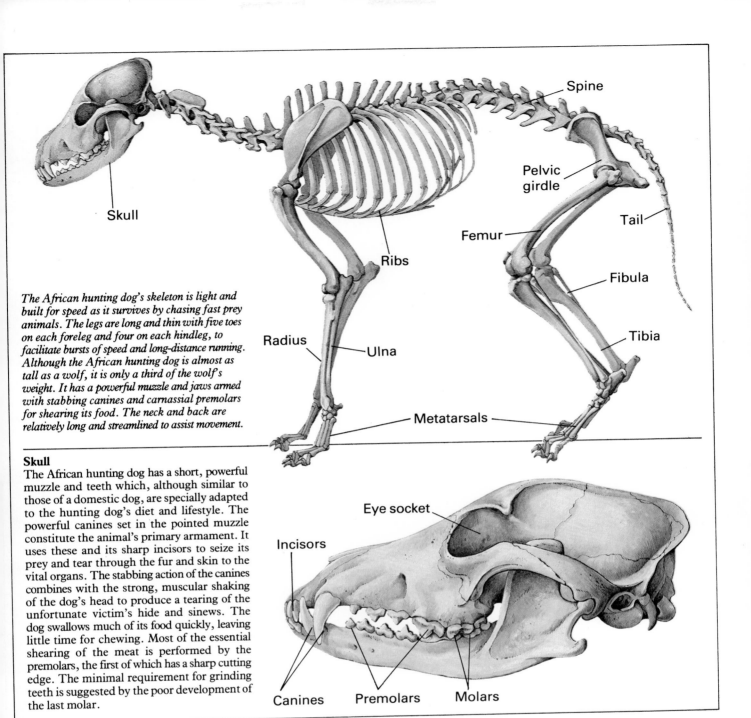

The African hunting dog's skeleton is light and built for speed as it survives by chasing fast prey animals. The legs are long and thin with five toes on each foreleg and four on each hindleg, to facilitate bursts of speed and long-distance running. Although the African hunting dog is almost as tall as a wolf, it is only a third of the wolf's weight. It has a powerful muzzle and jaws armed with stabbing canines and carnassial premolars for shearing its food. The neck and back are relatively long and streamlined to assist movement.

Skull

The African hunting dog has a short, powerful muzzle and teeth which, although similar to those of a domestic dog, are specially adapted to the hunting dog's diet and lifestyle. The powerful canines set in the pointed muzzle constitute the animal's primary armament. It uses these and its sharp incisors to seize its prey and tear through the fur and skin to the vital organs. The stabbing action of the canines combines with the strong, muscular shaking of the dog's head to produce a tearing of the unfortunate victim's hide and sinews. The dog swallows much of its food quickly, leaving little time for chewing. Most of the essential shearing of the meat is performed by the premolars, the first of which has a sharp cutting edge. The minimal requirement for grinding teeth is suggested by the poor development of the last molar.

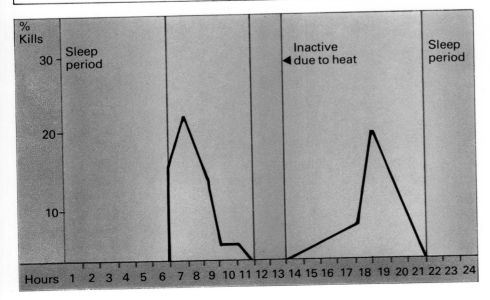

This chart outlines the activity of the African hunting dog — there are two main periods of activity in a usual day, peaking in the early morning and late evening. The pack hunts usually when the sun is low in the sky and the air is relatively cool. Around midday when the sun is at its hottest and beats down fiercely on the African savannah, the dogs enjoy their siesta period. Activity increases afterwards as the pack prepares for the evening hunt before it settles down to sleep after darkness falls.

The sequence of pictures above shows how a pack of African hunting dogs can successfully hunt and kill a large wildebeest. The leading dogs chase and tire the animal (1) while the remainder of the pack fans out to cut off its escape routes before closing in for the kill (2). They gather around the animal, nipping and biting at its head and hamstrings in order to bring the animal down (3). Before it can rise, the dogs start to tear open its flesh and eat it before it loses consciousness and dies (4). A whole animal can be eaten very rapidly indeed so that only the carcass and skeleton remain. The pack then returns to its base and the meat is regurgitated in a ritualised fashion for the young pups and the nursing females that did not hunt.

The diagram (right) shows exactly how the pack selects its victim and detaches it from the rest of the herd before chasing and killing it. On sighting prey, the lead dog breaks into a run and the others follow. The wildebeest is chased and headed off by the fastest dogs and is soon completely encircled, seized and then killed by the hungry pack.

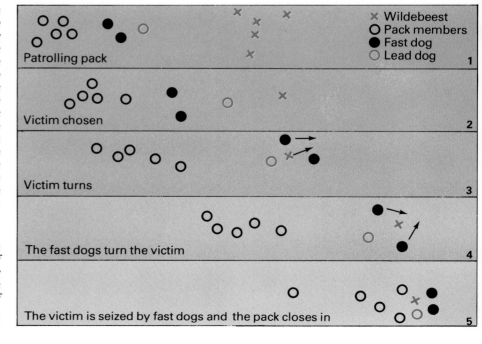

April	MORNING HUNT Animal killed	Weight in kgs	EVENING HUNT Animal killed	Weight in kgs	Kgs per dog per day
6			Warthog	53	4.8
7	Wildebeest	163	Wildebeest	40	18.4
8	Wildebeest (x2)	70			6.3
10	Zebra	219			19.1
11	Zebra	219	Wildebeest	45	24.0
12			Wildebeest	45	4.1
13	Wildebeest	45	Wildebeest	45	8.2
14	Wildebeest	30	Wildebeest (x2)	75	9.5
16					
17	Zebra	40			3.6
18	Thomson's gazelle	4			0.3
19					
20	Zebra	20			1.8
21	Thomson's gazelle	3	Zebra	219	20.2
22			Grant's gazelle	8	0.7
23			Warthog	86.5	7.8
26			Wildebeest	163	14.8
27	Thomson's gazelle		Wildebeest	45	5.5

This activity graph for the African hunting dog pinpoints the two main periods of activity during the day — early morning and late afternoon. During these times, the sun is low in the sky and the air is still cool and therefore they provide ideal conditions for patrolling the territory and hunting game. Activity becomes less pronounced in the late morning as it gets progressively hotter on the wide African savannahs and is non-existent at midday which is designated as a time for siesta. During the late afternoon, the dogs become more playful and active in readiness for the evening's hunt. A fully grown adult dog requires about 1.5kg of meat a day and a pack kills an average of one animal per week per dog.

The Gorilla

A mist swirled thinly through the trees, lingering in clinging eddies among the vines that festooned the dank branches. The upper boughs of the tree were bright with the notes of the dawn chorus but the rainforest floor was quiet save for the steady drip of moisture that slid from the canopy above. The trees grew close together and the undergrowth was rampant, green and wet.

The branches of a great *Musanga* tree quivered in the stillness and a shower of drops pattered down on the tangle below waking a full-grown male gorilla. His long hair was clotted in tufts and it steamed as his body heat evaporated the moisture clinging to it. He looked about himself carefully and ambled steadily in a shuffling gait to the nearest clump of bamboo and began to eat.

The mist was clearing and the warm sun struck through the thin foliage above the jungle floor. The tangle of saplings among the ferns began to stir and other black figures emerged. First, another large male, who gently pushed a four-year-old juvenile out of his way with the back of his hand, lumbered towards a sweep of succulent vines and began to tear off long strings of them. He wound them into a pad and sat down to bite off a piece. Occasionally he paused to pick out a dead leaf with his lips and to puff it away.

Within half an hour the little glade was populated by thirteen gorillas. All the adults were gathering food steadily and eating with quiet concentration. They had abandoned their shelters of small branches, which showed a darker green of crushed leaves surrounding the night's droppings. These were coarse-textured brown nodules of a stiff, fibrous material that would not stick to the gorillas' hair. The sun sucked up the morning mist to reveal the green forest clothing the steep side of an extinct volcano. Here and there the lava flows had resisted the decomposing work of the weather, and a bare patch of earth and rock showed.

The male who had been the first to rise looked around at the group of gorillas, hooted gently through pursed lips, and moved on feet and knuckles a few metres down the side of the mountain. He squatted by a bed of nettles, the fierce *Laportia alatipes*, which can sting through a layer of cloth. Without any sign of discomfort, he pulled a bunch of nettles. Pinching together a forefinger and thumb he drew the nettle stems through his fingers to strip off their leaves covered with stinging white hairs. He munched the wad of leaves with apparent enjoyment, sharing a few fallen leaves with a three-year-old who had bounded companionably up to him.

The other members of the group were scattered among the trees slowly choosing and eating their food. None strayed far from the large male, whose saddle of silvery hair made him conspicuous among them. The silver hair spread down from the nape of his neck to his hips. Above the heavy ridge of bone that hung over his brow was a wedge of silver hair which gave his general expression an air of concentration and thoughtfulness. The two other adult males in the group had not yet developed the silver-grey saddle that distinguished males over the age of fourteen years.

There were five adult females, two of whom were caring for young gorillas not yet mature enough to leave their mothers for long. One of the other females moved oddly towards a male juvenile of about four years of age. She held her left arm close to her body and walked on her feet and the knuckles of her right hand. She pulled a bunch

Opposite page:
Although lowland gorillas often live near water, they don't generally like it and are afraid to cross the smallest streams and drink little. They are found deep in the rainforests of Gabon, Zaire, the Central African Republic, south-eastern Nigeria and the Central African lakes region. The largest of the primates and the closest to man, the gorilla lives in small troops consisting of a single male, several females and their young. Each troop has its own small home range through which it wanders in search of food. Intelligent and fearless, the gorilla is a large ape, males being about 170cm high and weighing between 160 and 205kg in the wild, although they may be fatter in zoos. They have broad, muscular chests and shoulders and greyish-brown to jet black fur which may be silvery on the backs of males.

These mountain gorillas, found only in the Mount Kahuzi and Virunga Volcanoes regions, are feeding on a banana tree, stripping off the stem and eating the marrow. They often raid banana plantations in search of food and are persecuted by farmers sometimes for this reason. They are different in several respects from the lowland western gorillas, having darker, more silky fur, shorter arms and more man-like feet. They are found in open forests with plentiful food and abundant plant life. There is no defined birth or mating season and the young are born after a gestation period of 255 days. They develop quite quickly and are extremely playful and boisterous.

from the juvenile's shoulder hair and gave him a friendly cuff. Her left forearm was swollen and the long hair clotted with dried blood and pus. Nine days earlier she had stumbled while descending a steep bank. As she pushed out her arms to recover, her left arm caught in a knotted root and she broke her radius bone. Its broken edge pierced the skin. The injury had left her weakened as pus formed in the open wound, but for the last few days the flesh was drier and there had been less discharge. The hand and forearm would never again be useful to her and would be a constant source of pain and danger to her, but the wound seemed to be healing.

The nursing mothers carried their infants on their backs, the small creatures clinging to their mothers' black hair and staring inquisitively at the other animals. Occasionally, one would reach out a small hand, clinging tightly with the other, to pull a tender shoot. Holding it delicately between its lips, the infant experimentally nibbled on it.

Just before midday, having fed fairly intensively for more than two hours, the silverback male settled down on a fallen tree trunk. He raised his head and hooted gently, glancing round at the group that was now gathered in an area of about 160 square metres. The infants, juveniles and females made themselves comfortable quite close to one another. The two blackbacked males were outside the edges of the group but were close by. The gorillas settled down to rest. The only sound came from the activity of their digestive juices at work on the morning meal. The rumbles from their large bellies could be heard several metres away. One female squatted in the lower branches of a tree, her left leg dangling and swinging to and fro. The other adults dozed peacefully on the ground, with the exception of the injured female who nursed her arm restlessly. The juveniles and infants were not still for long.

A male juvenile of about three years old climbed onto a large fallen log. He sat there, now and then thumping the log with his

Although it is not normally savage and has a reputation for being a peaceful, gentle giant, the gorilla can be very frightening if attacked and may charge and cause terrible wounds. When gorilla troops encounter one another in the wild, the male leaders traditionally threaten each other, beating their chests and roaring as they rise on to their hindlegs. They will sometimes tear up handfuls of vegetation, tossing it into the air, and even make a bluff sideways charge. These displays are fairly frequent and are often a signal that the males do not want to fight. They therefore go through the motions of a mock show of aggression.

cupped hand and looking challengingly at the other young gorillas. One of these walked casually past the log, suddenly taking a leap up the clinging vines that ran up its sides. He nearly made it to the top but was bowled over by a playful push. Soon the others were joining in, and a game rather like 'I'm the king of the castle' developed as the young gorillas bounded up the log only to be toppled back again by the boisterous three-year-old on the summit. Finally, the 'king' hung on to two of his assailants and, locked together, they rolled to the ground. The resulting melée was dispersed after some shrill hooting from the nearest female.

The same three-year-old led the others off in a game of 'follow the leader' through the undergrowth, over the recumbent forms of the resting adults – who bore the disturbance with patience for a while – and through the branches of the trees. A harsh grunt from one of the nursing females warned the players of her annoyance as one stepped on her infant, and the game settled down to a more decorous pace for a time until the gorillas tired of it.

The rest period had lasted for two and a half hours when the silverbacked male rose on all fours. He walked carefully round the nearest nursing mother, who lay on her side, her arm curved round to hold her inquisitive youngster. The group's leader sought out a clump of bamboo. Hardly had he reached for some tender shoots when a strange silverback appeared on the other side of the rest area. All movement in the group ceased; hands froze in the act of carrying greenstuff to mouths. The young were the first to move after recovering from their surprise. They grabbed the hair on their mothers' rumps and hoisted themselves onto their backs, clinging tight. The group clustered together in a huddle a little to the rear of their leader. The great silverback, half as big again as other gorillas in his group, stood alone, staring at the interloper.

The leader's brow furrowed with the intensity of his stare, and his lips parted slightly, although his teeth remained hidden. The stranger, also on all fours, stared back. The tension grew as neither giant would give way. The stranger jerked his head to one side and snapped his teeth together. He turned sharply to present one side to the group's leader and, bending his arms outwards at the elbow to display their full length, he took a few strutting paces at right angles to a line between the two gorillas. All the time he watched his formidable opponent out of the corners of his eyes, a touch of white showing at the corner of his brown irises. A few seconds later, he turned and stood still, facing him again.

The leader had pursed his lips and begun to hoot softly. The hoots became louder and more rapid. He reached out a huge arm and wrenched a handful of stems from a vine and stuffed them into his mouth and then, blowing them clear, began to hoot again, louder and louder. His hair bristled, making his already colossal bulk yet more intimidating. With two stamping steps forwards he rose up on his legs to stand nearly two metres tall, his arms swinging in wide arcs to sweep branches and leaves high in the air. The noise was terrifying. The jungle seemed to shrink from this awful rage. He began to swing his cupped hands into his chest, beating it to make great thudding sounds, while his hooting had grown to a scream. His lips were drawn back from the huge canine teeth, which glowed with a reddish tartar. The scream changed to a deep echoing roar and he ran several metres sideways.

The stranger turned his head away from the group's leader but did not otherwise move until there was an earth-shaking thump as the great silverback smashed his open hand into the ground and charged. The stranger turned and fled before this hair-raising demonstration of ferocity. The leader did not pursue the interloper far but paused before turning back to his group, panting heavily. He had succeeded in showing off the stranger in no more than forty seconds without striking a blow.

It was only a short time before the group's search for food began again and tranquillity was restored. The afternoon passed quietly for the gorillas. They covered a distance of about a kilometre. They passed the freshly killed body of a small buck, half-eaten by a leopard, but the vegetarian gorillas left the body untouched. In the late afternoon, as the light faded, they came to a small river and turned to follow its bank upstream. For a few yards they followed the bank, showing signs of nervousness, until they encountered a fallen tree that spanned the water. The silverback led the way across to the opposite bank and into a stand of *Musanga* trees. The last gorilla to cross the river was a nursing female. Distracted by the hesitation of the juvenile in front of her, she slipped on a patch of peeling bark and she whined. The group turned to her, but she recovered and made the crossing in safety. The gorillas seemed more confident when the water was well behind them. Only one had paused to scoop up a handful of water and, holding it above her head, poured it into her mouth.

Once among the trees again, the gorillas began to build bivouacs for the night. Most of the group chose to sleep on the ground but one of the females without young climbed slowly into the boughs of a tree until she was about eight metres from the ground. Bending three branches down towards the bole of the tree, she trapped them against the main stem to make a bunch of foliage which she squashed into a comfortable shape.

The other gorillas gathered saplings and small branches for their beds. Some sat down beside vines and simply pulled them down to

This gorilla is eating grass which forms a staple part of its diet together with roots, stems, fruit and bark. Whereas lowland gorillas tend to eat a lot of leaves and fruit, their mountain cousins consume more bark. Volcano gorillas will often venture into the bamboo zone when the tender, young shoots are sprouting. The type of food thus varies according to habitat, but gorillas have never been observed to eat animal matter, eggs or insects in the wild. They usually feed in family groups, foraging with concentration for several hours and then resting while they digest the coarse fibre in their diet. However, as agricultural techniques improve and deforestation continues in central Africa the gorillas' natural habitat is seriously threatened and their main source of food could be substantially reduced.

ind around themselves in a saucer shape. The rims of these nests ere about 50cm high and the depressions in which the animals ested were about 35cm deep. The gorillas made little attempt to ne the nests, and one backblacked male chose so sloping a site that ardly had he settled down to rest than he began sliding down the cline. By morning he would be likely to have moved three metres r so.

The nest building on the ground took only half a minute, but the male who built a tree nest took four minutes to complete hers. The veniles, too, settled in their own nests, but the young infants, nder two years, stayed with their mothers. The group settled to eep as soon as the nests were complete. Soon there was little to be eard but the rumbling of the gorillas' bellies as they digested the oarse vegetation they had eaten.

The night air was disturbed by the sound of a lone male thumping is chest, and the great silverback rolled to a sitting position to give is own broad chest a few hefty thumps in response but, after eepily scratching his belly, he lounged back, sighed and slept.

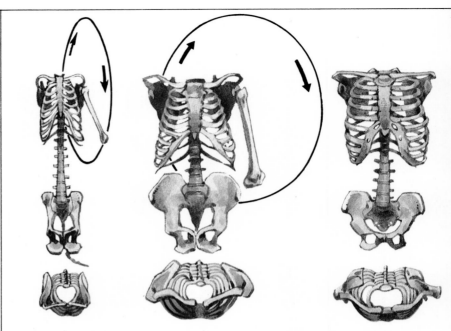

The skeletons of a monkey, a gorilla and a man are shown above. Although they are very similar in their structure, there are fundamental differences in flexibility and movement. For instance, a monkey cannot move its upper limbs far to either side of it, whereas the gorilla has greater freedom of movement, and the man can move his arms freely. His pelvic girdle is smaller proportionally than that of the gorilla.

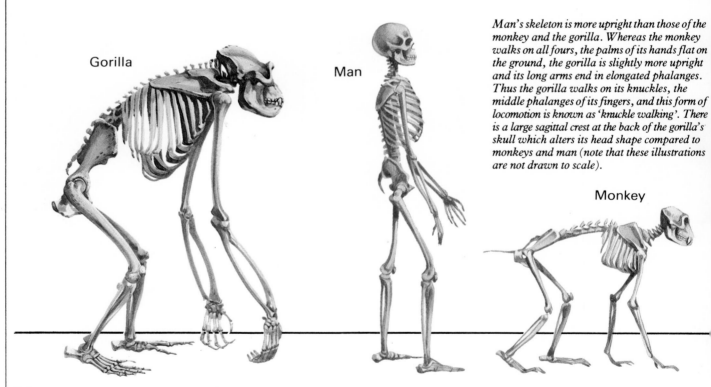

Man's skeleton is more upright than those of the monkey and the gorilla. Whereas the monkey walks on all fours, the palms of its hands flat on the ground, the gorilla is slightly more upright and its long arms end in elongated phalanges. Thus the gorilla walks on its knuckles, the middle phalanges of its fingers, and this form of locomotion is known as 'knuckle walking'. There is a large sagittal crest at the back of the gorilla's skull which alters its head shape compared to monkeys and man (note that these illustrations are not drawn to scale).

Primate posture

The three methods of locomotion found among primates are closely related to the structure of their skeletons. The monkey must climb trees rapidly and be able to run along their branches. To this end it has a long flexible spine with six or seven vertebrae between the pelvis and the part of the spine that is attached to the ribs. Its shoulder blades lie along the sides of the rib cage to allow free backwards and forwards movement essential to a four-footed mover, but it has little capacity for sideways movement.

The gorilla is a swinger, hanging from branches rather than running along them. This method of moving through the trees enables it to spread its great weight between two or three branches at once rather than trusting itself to one branch alone. The gorilla's shoulder blades are further round t the back, and allow for more comple movement through all planes. Its longer arm make swinging through the trees easier tha it is for monkeys, but on the ground, where i spends most of its time, the gorilla canno walk with its hands flat on the ground. Th length of its arms makes it adopt a half-uprigh position, supported on the knuckles of it

The legs of a gorilla and of a man are shown here. Whereas a man's legs are well adapted to walking in an upright position, the muscles being anchored by an upright pelvis and spine, the gorilla is capable of only a shuffling, half-crouching gait despite the strength of its leg muscles, necessarily powerful to mobilise its great weight. Its gluteus maximus muscles are comparatively weakly developed, and the gorilla cannot brace its knees in the same way as can a man. Consequently, the gorilla is able to adopt only a tiring position with both knees bent when standing on its hindlegs. However, its powerfully developed shoulder muscles enable it to swing from branches as it moves through the trees.

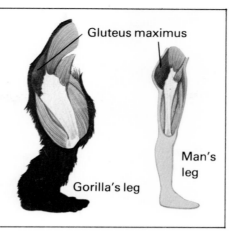

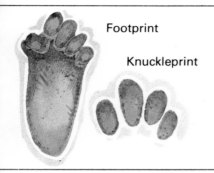

The gorilla usually walks on all fours and raises itself onto its hindlegs only in an aggressive posture. Its feet are plantigrade (adapted for walking on the whole surface of the sole) but the backs of the phalanges, the middle joints of the fingers, are used to support the weight at the front of its body. The backs of the phalanges are protected by thick pads of protective tissue. The footprint and knuckle-print of an adult fully grown gorilla are shown here (left).

ands. The gorilla has fewer lumbar vertebrae than the monkey, making its spine more rigid.

Man's more flexible spine and more upright pelvis are well adapted to anchor the muscles used for bipedal stance and walking. While a man can brachiate (move by swinging from branches), the weaker structure of his collarbones and shoulders make him far less efficient than the gorilla at this mode of travel.

Gorilla movement
In its first few weeks the gorilla's movements are uncoordinated. It in unable to hold onto its mother's chest for longer than two seconds.

From four to five weeks old the gorilla appears to focus on movements at close range and it can hold onto its mother for up to 10 seconds. It starts to look around and can ride unsupported on its mother's back at six to seven weeks, and three weeks later it begins to chew plants to supplement its diet of milk, and the process of weaning starts. After 16 weeks it learns to walk on all fours and to climb trees and beat its chest. It is not long before it can stand on two feet and starts to play with the other young gorillas. However, it stays with its mother until it is about three years old.

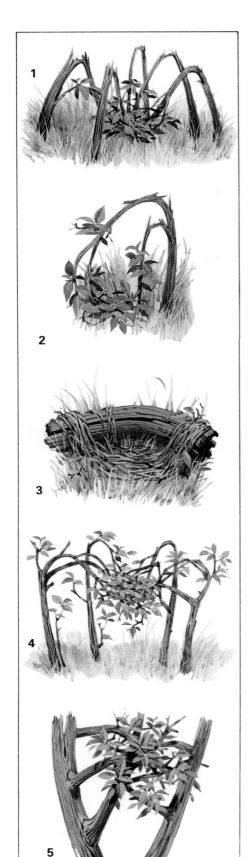

Gorillas build temporary nests for sleeping which are abandoned next day. To do this, they may bend saplings downwards (1); or break a small tree (2); swathe a fallen trunk with lianas (3); lock together several branches to form a nest above ground (4); or wind together branches in the bole of a tree (5).

At a glance
Mountain Gorilla
Gorilla gorilla beringei

Class	Mammalia
Order	Primates
Family	Pongidae
Height	1.25-1.75m on hindlegs.
Weight	Male: 140-200kg in wild.
	Female: 75-120kg in wild.
Weight at birth	About 2kg.
Gestation period	251-289 days. Single births are the rule but twins have been recorded.
Life span	About 30 years.
Diet	Exclusively vegetarian.
Coat	Black, sometimes with a russet tinge on the crown. All visible skin is black.

The gorilla clan (left) stays in a fairly tight group when feeding on the dense vegetation that forms a staple part of their herbivorous diet. They may forage with concentration for several hours at a time, eating vast quantities of plants and sometimes leaves or fruit. For mountain gorillas, such as these, food is plentiful in the high forests where there is abundant ground level vegetation. Lowland gorillas tend to eat more fruit than their mountain contemporaries. Some essential minerals are added to their diet by eating a little earth from mineral scrapes which the gorillas visit from time to time. However, they obtain all their energy from plant nourishment, eating the stems, roots, flowers and pith of the jungle plants. After feeding, the gorillas rest for a while to digest the coarse fibre.

This diagram indicates the feeding route of a troop of gorillas as they move through the central African forests in search of food. They rarely move more than 1800m per day when feeding. This map of their movements over a period of 16 days is based on the observations of the naturalist George Schaller who studied gorilla troops in the Virunga Volcanoes region for two years. The factors that affect the distance travelled by the group seem to be the density of the available fodder and the state of the weather. Evening nesting sites are marked by dots on the map. The gorillas double back and cross their original route many times as they appear to wander aimlessly through their defined territory. There is no distinct pattern in the group's wanderings and wherever they happen to be at nightfall they build their nests with whatever material is available. Mountain gorillas usually build them on the ground.

Gorillas are found only in western and central Africa. This map shows the present ranges of the lowland and mountain gorillas. The western, or lowland, gorilla inhabits the hot equatorial forests of Zaire, Gabon, Equatorial Guinea and the south-western regions of the Central African Replublic as well as the extreme south-east corner of Nigeria. However, the mountain gorilla is found in the montane rainforests and bamboo groves of the mountain slopes in the Virunga Volcanoes region between Lake Tanganyika and Lake Albert. There are no gorillas in the lowland forest area lying between these two ranges. Some naturalists believe that gorillas once ranged from the Cross River in the west to the northern reaches of the Ubangui River and then south to Lake Albert. However, they are now extinct in these areas and the reasons for their disappearance are still a mystery to scientists.

The Hippopotamus

The air was heavy with moisture, although the dawn had lightened the sky four hours earlier. Grey clouds showed a soft blue distance between their ballooning clusters. Out over the plain, distant rain fell heavily in precise parallels just as it had done for several weeks. Here, below the waterfall, a patch of sunlight warmed the pools of the swamp that lay half a kilometre to the east of the main river. To the north, the broad river fell in a white sheet over the rocks of a low scarp edge. The deep pool at the foot of the cascade slowed the river to a brown-stained, stately flow past steep banks. The west bank had little vegetation on its sides, just a scatter of low bushes. On the east bank there rose a low mound of acacia thicket, around which a slow trickle drained from the swamp beyond into the river.

In the swamp and along the outflow to the river, papyrus reeds reared their graceful plumes seven or eight metres above the muddy ground. Below them, hiding much of the waterlogged surface, a dense forest of matatite reeds rippled here and there. The sunlight glistened on the jewelled plumage of a malachite kingfisher as he perched on a bent stem of papyrus and casually watched the waters for movement. He was not alert, knowing that he was unlikely to find fish in these waters where the rafts of water hyacinths grew thickly. His head turned to watch a dull brown hammerhead stork land with much flapping on a brown patch in a swamp pool.

The bird cocked its head to one side, then drove its strong bill into the muddy surface near its feet, pecking and tugging in a businesslike way. It threw its bill upright and swallowed. At the same moment it tilted sideways and flapped half-opened wings to

Hippopotamuses love to wallow in muddy pools and rivers during the heat of the day, often with only their eyes, nostrils and ears showing above the surface of the water. These sense organs are all located on top of the hippo's head so that the hippo can remain partially submerged for long periods, thus becoming weightless and sparing its short legs the effort of supporting its huge, heavily built barrel of a body. Its name literally means 'river-horse' and it is well adapted to its semi-aquatic life. Although it is usually to be found on sandbars and open river banks, it retreats to the cover of reed beds in areas where it is persecuted.

keep its balance as its foothold swung about. The bird was unperturbed as the head of a hippopotamus rose from the waters and the mighty beast lumbered to a fresh position.

Its back and head were festooned with water weeds and a bright pink lily slid from one of its ears, which was vigorously shaking drops of water from its deep central fold. Both ears were oscillating in an extraordinary manner. She settled to rest again, nudging another hippopotamus gently, and resting her head against the other's side. The pool was quiet again except for the sighs of slow breathing that moved the still, warm air. Every minute or so a hippopotamus would raise its head slightly, the nostrils opening to blow out air with a fine spray of moisture, accompanied by a snorting grunt. Each of the dozing beasts raised its head for a snoring change of breath about once every two or two-and-a-half-minutes.

One hippopotamus butted another, which rolled slowly onto one side. Only one nostril and one closed eye was visible above the water but the animal was undisturbed. She raised her head sideways from the water when she needed to breathe and sank down again, still lying on her side. Next to her, a baby hippopotamus's nose broke the surface, breathed and slid below again with scarcely a ripple.

Beyond the pool, about 30m from the herd of females and juvenile males, a large bull hippopotamus rolled onto his back in a deep wallow. The mud encrusted his body, protecting it from the sun. Even when the light was not fierce as it was in the hot season, he could lose 18 percent of his skin moisture in a day. This could make his skin dry and cracked so, instinctively, he protected himself with mud when away from water. The mud coat helped to reduce the irritation from skin parasites too. He rolled and grunted, slopping about in the wallow with evident enjoyment.

From a wet gully that led to the swamp's outflow, sounds of two more bulls suggested their irritability. One trotted to the edge of the pool and plunged in. Another followed but turned away when he saw the big bull leave his wallow and walk threateningly towards

Although they were once a common feature of African rivers, hippopotamuses are now found only in the area north of the Zambezi River to Khartoum in the Sudan, apart from wildlife reserves and national game parks. They live in highly organised social groups of between 20 and 100 members known as schools. Each school has its clearly defined living quarters and feeding territory. The females and young live in the centre, with the juveniles and adult males in 'refuges' around the perimeter of the central creche. There are special paths leading to the feeding grounds. Even the most aggressive male is cautious when entering the exclusively female creche because he will be attacked by all the females present if he fails to be deferential and observe certain rituals. For example, if a female rises he must sit down.

the pool. With a sequence of squelches, he entered the water and approached the young male. The youngster took a careful look at the bull's gaping mouth with its enormous canines exposed in the lower jaw and he turned away. As he turned, he defecated in the water, heaving his hindquarters out of the water and swishing the matter across the pool in front of the approaching bull. The bull paused and watched the young male scramble up the bank to trot away towards the river. Its action as it fled from the dominant male left the bull satisfied that he was still the acknowledged chief.

Just before midday, the females stirred themselves. For some time, two baby hippopotamuses had been plaguing their mothers, climbing out of the water to visit the big male in his wallow, only to be shooed back again by a female. Now, the herd slowly made its way down to the mudbanks that stretched out into the river where the swamp's outflow entered it. The pair of young walked close to their mothers. Whenever they showed signs of straying out of touch, the mother butted the youngster back into place.

The water ran in a deep channel between the bank and the ridges of silt that rose above the water's surface a few metres further out. The current was slow, folding back on itself in a gentle eddy. The small calves soon disappeared from view under the surface, their short legs pushing the riverbed and paddling forwards towards the banks. The females bumped them back on course when the light flow of water edged their young away from the safety of their mothers' sides. Every 20 seconds or so the youngsters pushed themselves to the surface to breathe but some of the large females strolled along the riverbed for four or five minutes before surfacing.

As the riverbed rose near the mudbanks, the calves sported about, watched by their mothers and aunts. Most of the females dragged themselves up the slopes of mud and they waddled their rumps into the soft silty ground before rolling onto their sides. One of the calves struggled out of the water to suckle at one of his mother's two nipples. The sun warmed her and she gave a great sigh. The loud hiss of air escaping through her nostrils failed to alarm an egret

Although hippos give the impression of being slow and ungainly on land, they are really surprisingly fast and agile and can overtake a man running. The second largest land animals, they measure about three and a half metres long and may weigh up to four tonnes. Their huge, barrel-shaped bodies rest on squat legs and are hairless apart from the odd bristle on their tails and muzzles. They are often said to 'sweat blood' owing to the pink, sticky fluid secreted by glands below their skin for lubrication. This may also have antiseptic properties as the most horrifying, gaping wounds received in fights heal quickly and cleanly without suppurating.

which was strutting over her wet shoulder to pick daintily at the hippopotamus's ear. The ear rotated rapidly but the bird timed the swift movements accurately and relieved the female of another parasite.

The second calf walked to the highest part of the banks where a good-sized young bull had defecated. He was flicking the droppings, which were rather like those of a horse but softer, with his short, flattened tail. The youngster watched him from close by. As the male squelched his way further down the bank, the calf sniffed at the droppings for a few moments before stumbling in the wake of the larger beast. He sniffed at the male's rear end and gave it a few licks as if to memorise the taste and odour of this important member of the herd.

The herd was a fairly large one. There were seven females, four of them capable of bearing calves and the others between five and eight years old. The eight-year-old was soon to be sexually mature. The herd had six males. Two of them were still immature – one five-year-old and one on the verge of puberty at seven. The males scattered themselves at some distance from the females, which stayed close to each other. Only the big, dominant male positioned himself near the females so that he could interfere with any attempt on the part of other males to challenge his authority.

Sunlight faded early as rain soaked down on the peace of the mudbanks. Most of the hippopotamuses slid into the water. The adults moved out to dawdle their feet lightly through the mud of the riverbed in about one-and-a-half metres of water. Their heads scarcely broke the surface. Eyes, nostrils and ears only pricked above the oily waters that hissed with the falling rain.

As the riverbed swirled with silt stirred by the hippopotamuses, nutrients spread through the water. They were followed by a shoal of fish, looking a little like carp. They appeared to be quite unafraid of the great beasts slowly moving through their element. They swam up to them and picked at fragments of algae adhering to their skins.

It was quite late in the afternoon when the rain drifted away from the river. As the sun caught the distant white fall of water upstream, a solitary hippopotamus ambled into view, walking slowly through the shallows. As she walked nearer to the herd, the big bull swam to meet her. He noticed a young calf at her side, walking with the side of her body pressed against her mother's. The bull turned away to swim towards one of the other males of the herd, which also wanted to investigate the female. The old bull opened his huge mouth and bellowed. He swam quickly towards the smaller male, keeping his mouth open to display the yellowed canines. One of them had been bent sideways in a battle the previous year. Although it no longer sharpened itself against the shorter canine in the upper jaw whenever he closed his mouth, it was still sharp enough to rend an opponent's hide.

The young male submitted to the superior force of the bull and stirred the water into foam as he turned away and made his best speed into the middle of the river. The bull submerged to stroll along the riverbed towards his herd. He surfaced a few metres from the basking hippopotamus and remained there buoyed up by the brown water, his feet exerting hardly any pressure on the bottom.

The female and her calf nosed their way among the rest of the herd, greeting them with sniffings and rubbings of heads and sides. She had wandered from the herd a few weeks earlier to have her calf alone. Now, returning, she soon found herself a part of the group again. She left her calf for a short time while she chewed at some

This hippo is yawning with its enormous mouth opened wide — a characteristic gesture which has nothing to do with feeling tired. Instead it is an aggressive challenge to fight. The threatened opponent can only avoid a fierce fight by lowering its head in a submissive posture. Territorial fights are both common and dangerous, with the contestants rearing up out of the water and attempting to slash each others' bodies with their long canine tusks which are used only for fighting. They are lethal weapons, which grow continuously and are kept razor-sharp by constant wear against the other teeth. Although wounds, however ghastly they may appear, usually heal fast, a hippo may starve to death if its foreleg is broken as it cannot leave the water to feed on grass.

This hippo is totally submerged on a riverbed and can walk along the bottom, feeding on succulent waterplants, by trimming its buoyancy. With its nostrils tightly closed, a hippo can remain completely submerged beneath the water for up to five minutes at a time. It surfaces eventually with a loud snort, which can be heard some distance away, as it expels the air from its lungs.

water cabbage which floated on the still waters of the small delta where the swamp outflow ran into the river. The small calf tried to follow her across the deeper water that flowed between the mudbanks and the riverbank but when he swam near to the pull of the gentle current, a female of the herd swam after him and nudged him back to the safety of the shallows. She stayed close to the little animal, licking her and nuzzling her until her mother returned. The calf sank down and briefly sucked her mother's milk underwater.

The twilight was short. The clouds dimmed the sunlight before its source sank below the horizon. The young males made their way to the west bank of the river. The banks were steep there but three brown scars in the bank, each a few score metres from the other, led to gullies offering an easier gradient to the top. The young males scrambled up the bank and entered the gullies. These were not natural grooves in the bank but hippopotamus-made. They were narrow at the bottom and wider where the animals' flanks had polished the mud walls of the gullies smooth. The hippopotamuses placed their feet close together as they mounted the narrow path to the top of the bank. They paused to mark the route with droppings. Near the top of each of the paths were piles of droppings, of which one pile was a metre high.

The old bull lurched from the water with a loud grunt and began to climb after the younger males. His scored hide brushed against the sides of the pathway. The scars of many successful battles puckered the thick skin along his sides and there was a series of ridges high on the shoulder where a lion many years earlier had seized him. He had managed to roll down the steep bank into the river with the lion hanging on. Once in the water, he had rolled the terrifying beast under until it released him. He had slashed it along the ribs with his canine teeth but the wounded lion had scrambled up the bank and slunk away roaring furiously, while two lionesses, which had tried to help in the hunt, followed him at a distance. The hippopotamus had been a youngster then and not fully grown. It would have been a desperate pride of lions that attempted to bring him down now that he was at his full strength.

The herd joined together where the paths leading from the river met. They clopped their way along the paths in the dim light, making their way towards the short grass that they liked best. As the main path wound along, worn deep with use, the hippopotamuses had to step over the gnarled roots of trees that had once grown there. The path had eroded the soil around the old roots, which now twisted their way across it, forcing the animals to step over them carefully. They followed their road for a little over three kilometres before the sides of the path grew lower and the animals could spread out to graze the grass and herbage around them.

They bent their massive heads to the ground and pulled at the grass with their horny rimmed lips. Occasionally, they used their incisors to tear at a clump of succulent herbs but most of the time they grazed the short rich grass, cropping round the well-grown tussocks. They were quiet. The only noise was of tearing grass and munching molars, punctuated by a few grunts and blowing noises.

The night was a dark one. The stars of the savannah were obscured with cloud and it began to rain again. Perhaps it was the rain and the blackness that caused the old bull to see the stranger only when he was at close quarters. Neither animal had been aware of the other until an eddy of wind had cleared the clouds for a moment from the moon. In the sudden hard light, metallic gleams were reflected from the bushes and the wet hides of the grazing animals.

Baby hippos are well instructed by their mothers in the rules of the hippo community and learn by imitation. They learn to walk at their mothers' sides, well protected from aggressive adult males and are punished with a blow of the head or slashing great tusks if they are disobedient. If the mother leaves the creche to feed or mate, she entrusts her baby to another female in her absence and babysitting is common. The young hippos are born after a gestation period of about eight months and within five minutes of birth the calf can swim, walk and even run. It is suckled by the mother for about one year although it does not reach sexual maturity until it is eight or ten years old. The young hippos play together, rolling in the muddy water and indulging in mock fights.

The stranger was walking directly towards the herd's dominant male. For a moment, he stood still as if shocked at the impertinence. If the interloper had been a man or another sort of animal, he might well have turned and trotted back to the safety of the river but the other male, a fully mature, well-scarred bull like himself, was a different matter. He threw his head back, opened his mighty jaws and roared defiance. The stranger stopped in his tracks, then bellowed back.

For some time, the two eyed each other with their mouths open to display the terrible nature of their weapons. They stood about 50m apart, roaring while the rest of the herd trotted back towards the river. The first to charge was the stranger but he began by only a few seconds' margin before the herd bull ran forward too. They moved with a speed that was surprising for such huge animals and as they passed each other, they swung their heavy heads to one side and slashed at each other's passing flanks with their canines. They passed, turned and charged again. The stranger was bleeding heavily from a long rip in his side. Once again they passed, slashing sideways. This time both animals staggered but recovered their balance and turned to face each other again.

The stranger's mouth was pouring with blood and one of his canines had gone. The whole tooth seemed to have disappeared. The old herd bull swung his head and roared. Protruding from his side, well back along his flank was the stump of the stranger's great tooth. Nearly 20cm of it was embedded in the bull's side. For some minutes the two giants gathered their strength for another charge.

Although hippos spend most of their daylight hours wallowing in mud like these ones, they may wander many kilometres at night along grazing zones to feed on shoots of grass. The width of these zones varies according to the seasons, decreasing in size when the grass is long and lush in the wet season and increasing in width in the more arid, dry season as the hippos have to travel further in search of food. A hippo can eat 70kg of grass in a single night as it sways along established trails, lined by piles of dung to act as scent markers in the dark.

The tactics did not change – again and again they passed flank to flank and ripped at the other. The old bull, terribly wounded in the second pass, now had the better of his antagonist, who rarely managed to hit with his surviving tooth. For over two hours the great bulls fought. The periods between their charges grew longer and longer as their energy failed but still neither would leave the field.

The clouds gathered round the moon, blinding its light at the moment when the exhausted animals thundered towards each other again. Weary as they were they charged again, the old herd bull instinctively slashing sideways and upwards. He was lucky. His right canine, the one that had been bent out of position years before, caught the stranger under the ribs. As they rolled together in the rutted earth, the tooth held and drove deep into the younger hippopotamus's heart.

The pair dragged themselves apart. The herd bull reeled to his feet but the stranger could not make it. He sank with a bellow to his belly, his legs refusing to lift him. Slowly, he rolled over onto his side, dying. The herd bull uttered a great bellow and walked slowly down the path to the river.

He paused often, his head hanging low. Even when he bellowed, he could not find the strength to lift the heavy head high. It took him until dawn to reach the steep descent to the river which had always been his home and refuge. For many minutes he swayed on his feet at the top of the slope. Then, staggering forward, he slumped to his knees and slithered down the bank to the water. As he rolled into the river his blood ran out along the current like a broad streamer. He raised his eyes and made a last, bubbling grunt as he slid into the deep water and died.

Downstream a little way the herd watched his body float past. One of the males and two females swam out to nose it gently for a while and then they returned to the mud banks. The largest of the males swam closer to the females and ground his rump into the mud before settling on his side.

The hippopotamus family

The family Hippopotamidae is confined to Africa; its early members inhabited the swamps, lakes and rivers as far north and east as Egypt and Palestine in past times, but by the early years of the last century their threat to crops of the fertile delta and the Jordan had been eliminated, and the bellow of the hippopotamus was heard no more in these lands. In the rest of Africa, the animal thrived in immense numbers well into the present century. The coming of the high-powered sporting rifle decreased the great herds of all the mammals of the continent, and the hippopotamus was one of the hardest hit. Barely credible numbers were shot and left to rot. The fine ivory of their teeth was the principal cause of the great beasts's undoing. When traders stripped the glass-hard dental enamel from the huge canines — the largest ever found was 64.5cm long — they exposed white ivory which never yellowed with age unlike the ivory of elephant tusks.

There are two genera of hippopotamuses; the larger animal, *Hippopotamus amphibius*, is the better known and more populous one. Most zoologists agree that there is only one species in each of these genera, but some have defined subspecies of *H. amphibius*. These are based on slight differences in the structure of the skulls. The species are *H.a. amphibius*; *H.a. kiboko* (Heller); *H.a. tschadensis* (Schwarz); *H.a. consrictus* (Miller) and *H.a. capensis* (Desmoulins).

The second genus, *Choeropsis*, is a rare beast with only one species, *C. liberensis*. It is commonly called the pygmy hippopotamus and is more pig-like than its larger relatives. The pygmy has a more terrestrial habit and lives singly or in pairs. It does not congregate in the big herds favoured by the larger hippopotamus. The pygmy's feet look much like a pig's trotters. The two posterior digits are placed quite high and all four separate well, making the animal better adapted to movement on land than in the water. Its body is less barrel-shaped and more like a torpedo than that of the *Hippopotamus*.

The **pygmy hippopotamus** (*Choeropsis liberensis*) stands 78-83cm high at the shoulder. It is about 150cm long from nose to rump and weighs 180-260kg. Its young are much smaller than their parents and smaller in proportion than the offspring of many mammals. At birth, a pygmy hippopotamus weighs 4.5-6.2kg. It is born after a pregnancy of 201-210 days. No one has observed the birth of one of these calves in the wild, but if births in zoos are to be taken as a guide, the pygmy gives birth on land, not in water as *H. amphibius* often does. The pygmy's head is rounder than the larger animal's, its back more arched and its rump more sloping. The skin is blackish, lightening a shade on the underparts. The pygmy has only one pair of incisors in the upper jaw. Its diet consists of grasses, vegetables, succulent plants and fallen fruit.

Hippopotamus (*Hippopotamus amphibius*) See *At a glance* panel on this page.

At a glance *Hippopotamus* *Hippopotamus amphibius*	
Class	Mammalia
Order	Artiodactyla (even-toed ungulates)
Suborder	Suiformes
Family	Hippopotamidae
Genus	Hippopotamus
Length	3.75 – 4.5m from nose to rump
Height	up to 1.65m
Length of tail	15 – 56cm. The tail is round in section
Weight	3 – 4 tonnes
Young	One calf of about 27 – 45kg is born.
Gestation	227 – 240 days
Diet	Herbivorous; mostly grasses with some aquatic plants and herbs.
Skin	Fine hairs with bristles around the muzzle and a tassle at the tip of the tail. The skin is brownish grey, slightly pinkish round the muzzle, eyes and throat.
Lifespan	In the wild, 30 – 40 years.

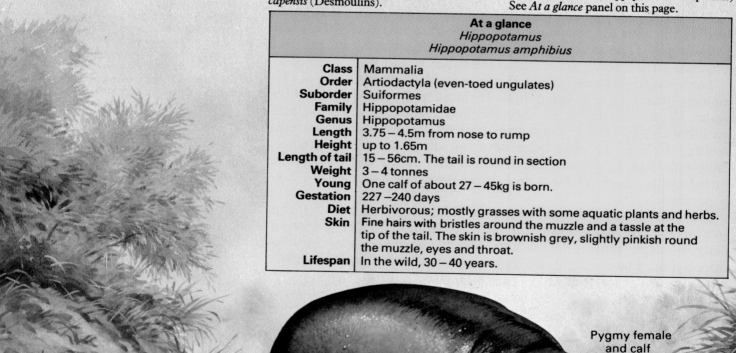

Pygmy female and calf

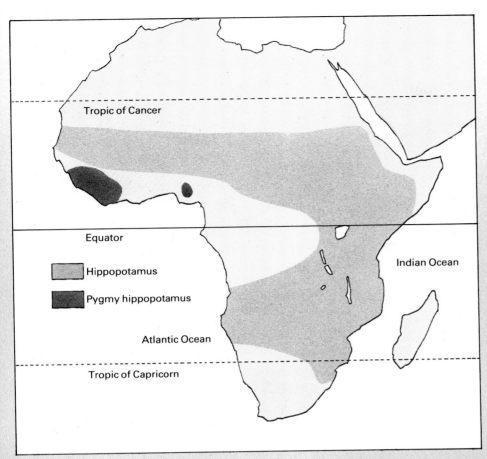

Whereas the herds of H. amphibius *are spread widely over the swamps, lakes and watering places of the savannahs of the African continent south of 17° latitude north, the rarer pygmy hippopotamus is confined to a relatively small area of dense forest and swamp in West Africa — Liberia, Sierra Leone and Nigeria. Hippo herds congregate especially around the great African river systems — the Nile, Niger, Congo, Zambezi and the Limpopo rivers and their tributaries.*

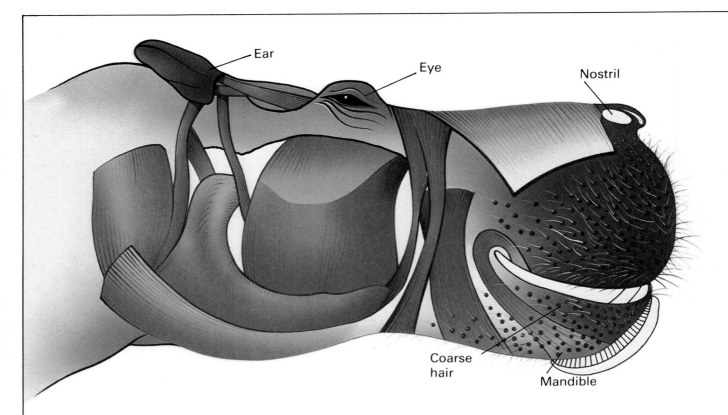

Lakes, rivers and swamps are home to the hippopotamus herds, which spend as much time in the water as they do on land. The water is both the hippopotamus's nursery, a place for relaxation and a refuge in danger. It is an animal well adapted to this watery habitat, but without upsetting its capacity for leading its life on land.

The bulk of the hippopotamus's body, easily supported in water where it treads lightly over muddy riverbeds and lake floors, is inconveniently very expensive on energy requirements on land. Strong, short legs enable the animal to climb even the steepest banks and when alarmed the hippopotamus can make surprisingly good speed towards the relative safety of its territorial waters.

Skin

The hippopotamus has a thick, tough skin which is not as naked as it may appear at first or even second glance. It has fine hairs that grow thinly from the follicles embedded in the skin. On its back, the animal has maximum protection from about 50.8mm of hide, but elsewhere the skin is thinner, although it still offers excellent security from any but the most violent assault. Within the animal's skin are the glands that produce the well-known 'blood-sweat'. This pinkish-red sweat does not occur continuously but only for a short period after the hippopotamus leaves the water. The glands that produce it lie a little less than 3mm below the skin; they are about 1.5mm long and 0.5mm wide. The glands have no limiting membrane so their product escapes easily through ducts to the surface. The so-called 'blood-sweat' consists of red corpuscles and a colourless liquid.

Teeth

The hippopotamus has only to open its mouth for its impressive armaments to become immediately obvious. The incisors are useful for scoring the earth so the animal can lick the minerals that it needs to enrich its diet. The canines are almost exclusively fighting weapons, growing to great lengths in the lower jaw.

This cut-away illustration of the hippo's skull shows the internal structure and muscles. The eyes, ears and nostrils — the principal sense organs — are all set high on the head for when the hippo is immersed in water. The broad lips (50cm wide) are ideal for plucking short grasses. A few sparse bristles grow on the muzzle of the head which is small in proportion to the rest of the massive body. The large mouth contains a formidable row of teeth on which the enamel gradually wears down with age.

Those in the upper jaw fit against the lower ones, sharpening them to knife-like edges. The premolars and molars function as choppers for reducing the tough grass and herbs of the hippopotamus's food to digestible wads.

A calf is born with ready erupted incisors which are shed after a few months and replaced with a set of permanent teeth. These have large pulp cavities and grow continuously, appearing yellowish because of the thick coating of protective dental enamel which gives the teeth a hard surface resistant to the severe

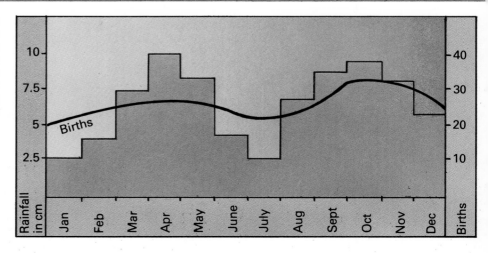

This chart shows the relationship between the number of hippopotamus births and the mean monthly rainfall in the Queen Elizabeth National Park, Uganda. The highest birthrate usually coincides with the wettest months, the numbers of births falling in the dry seasons when there is less food for the mothers and newborn calves. However, births occur throughout the year, peaking in April to May and in October.

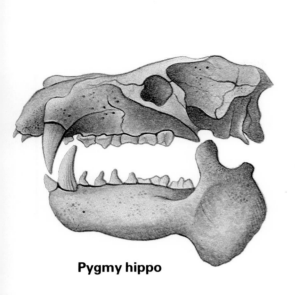

Pygmy hippo

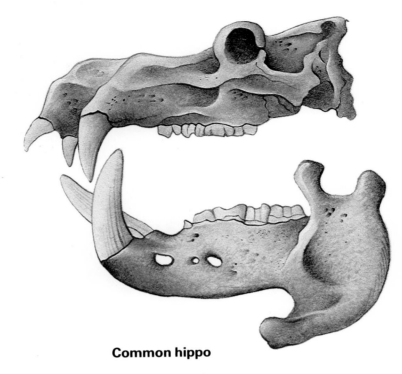

Common hippo

These diagrams show the differences between the skulls of the common and pygmy hippopotamuses. The head of the pygmy is rounder and has only one pair of incisors in the upper jaw unlike the common hippopotamus, whose lower jaw is larger and more powerfully constructed.

wear they must endure from the animal's coarse diet. After many years the enamel wears on the grinding teeth. The pattern of exposed dentine becomes greater with age, offering an important aid to zoologists in determining the age of a hippopotamus.

Stomach

The hippopotamus's stomach has a great deal of work to do. The animal does not chew its herbage well so the stomach has to accept a mass of crudely chopped roughage. Like ruminants, the hippopotamus, which is not a ruminant, has a stomach of four chambers, but unlike the ruminants the 'river-horse' does not chew cud. The stomach has to perform all the work of digestion in one operation.

The four chambers may be considered in three divisions. In the two anterior diverticula and the large median chamber, the process of breaking down the herbage takes place. In the posterior chamber, the stomach secretes gastric juices which activate the fermentation process. In this digestive system, the cycling of the food is slow. The food remains in the stomach for a long period before passing down the 50-60m of intestine.

The stomach of a male that has fed well contains about 34.9kg of herbage; this accounts for 12.8 per cent of a male hippopotamus's bodyweight. A female eats rather more: 15.2 per cent of her total weight. Although the intake of food is not as high as might be expected of so large an animal, it might be compensated for by the long periods spent in taking restful advantage of the water's buoyancy. Hippopotamuses do not rely on long bursts of speed when escaping from any predators so a large stomach that retains its fodder for a long time while digesting it efficiently is not a serious disadvantage.

Births

Female hippopotamuses may give birth at any month of the year, but the numbers of calves born peak at the rainy seasons. In Uganda, where the two scientists, R.M. Laws and G. Clough, correlated rainfall and births, these peaks occur in April or May and October.

In other parts of Africa, the rainy seasons fall in different months, and the rise in the birth curve coincides with the wettest months. The herbage in these wet months is protein-rich and makes good fodder for the nursing mothers.

Most females bare single calves. To give birth, they generally go down to shallow, still water. When they go into labour, they lie on their sides and the calf is born in a rush, followed by a jet of blood. The calf sinks for a few moments, then struggles to the surface for its first breath of air. The mother noses it towards a sandbar or a bank where she lies down so that the youngster can search out one of her two nipples and begin to feed. Unlike other ungulates, the mother hippopotamus does not eat the placenta.

Calves stand a fairly good chance of survival although about 20 per cent will die in the first year of their lives. Between one year old and 33 years only six per cent of the survivors will die, but as one might expect, the chances of survival become slimmer after about 38 years.

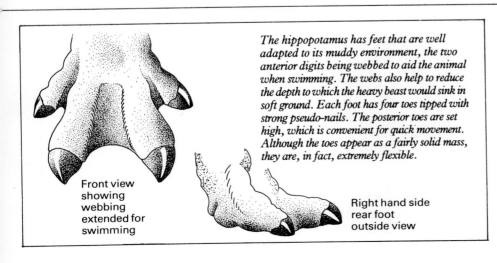

Front view showing webbing extended for swimming

Right hand side rear foot outside view

The hippopotamus has feet that are well adapted to its muddy environment, the two anterior digits being webbed to aid the animal when swimming. The webs also help to reduce the depth to which the heavy beast would sink in soft ground. Each foot has four toes tipped with strong pseudo-nails. The posterior toes are set high, which is convenient for quick movement. Although the toes appear as a fairly solid mass, they are, in fact, extremely flexible.

The Hummingbird

A row of dun coloured birds roosted, untidy in the dim green light of dawn. From these insignificant bundles of ruffled feathers, the tiny beaks pricked upwards towards the sky like delicate thorns. There was no sign of movement. Their bodies, no more than ten centimetres long – and some much smaller – were deathly still. All night they had roosted in a state that in other circumstances might have been called hibernation. All the forces that meant life to those fragile forms had slowed down in the chill of the night until the fresh light of the new day, warming them through, restored them to conscious life.

Their breasts began to move and flicker with life. Breath filled their lungs, the heartbeats became stronger and hunger stirred in their stomachs. They fluttered their plumage into order. First one, then another, stretched up on its legs, three toes of each foot extended forwards and one toe behind grasping the twig. Wings whirred, and they moved so fast that they disappeared into a soft blur. Their feet relaxed, disengaged their grip, and, without a spring or a step, the birds, one after another, rose smoothly and slowly into the air.

They darted quickly in various directions towards the gorgeous opening flowers that grew from the jungle floor and hung down from the branches of the trees. A sunburst of colour, threads of flashing green and blue and red shot across the shafts of sunlight that penetrated the canopy of foliage. What had appeared as brownish, dull-looking birds on the roosting branch were turned by the magic of light into small arrows of prismatic colour.

The birds approached the plants to feed. A ruby-topaz hummingbird hovered briefly before alighting on a flower stem and searching the corolla of the blossom for the nectar within. The blazing gold of his chin and the iridescent ruby of his head outdid the flower in richness. He pushed out his long tongue beyond the tip of his bill to seek the nectar in the flower's bell. His tongue was long and white. It pulsed rapidly in and out about ten times a second, gathering the food. The end of his tongue divided like a snake's, but the inner edges of the fork were lined with a fringe. The bird curled the forked edges over and flicked the moist nectar along its tongue to a point where two parallel horny grooves ran back. The nectar slid along the grooves until the ruby-topaz could swallow it.

The feeding operation happened at great speed. When the bird had taken the nectar from one flower it whirred into life again, its wings humming like an insect's. For a moment it hovered, barely clear of its perch, before moving sideways and settling down further along the stem to probe another flower. Some of the pollen sticking to his bill dusted off on this flower, fertilizing it. (Many of the flowers visited by hummingbirds depend almost entirely on the birds as pollinators and have evolved flowers that match the slenderness and length of the birds' bills. The species with short bills tend to feed from the shallow flowers, and those with long ones from blooms with deep corollas.)

The birds were attracted mostly by the red and yellow blooms. Anything that was bright red seemed to draw their interest at once. Unlike bees and other nectar gatherers, the hummingbirds did not

Opposite page:
This tiny hummingbird is hovering in mid-air as it extracts pollen from a colourful hibiscus flower with its long bill. The hummingbird's high metabolic rate necessitates regular feeding, and nectar forms the staple part of its diet. Its bill is specially adapted for feeding in tubular flowers, being slender and slightly curved to penetrate the blossoms. The elongated tongue can often be extended beyond the beak and the edges curved inwards to form a hollow tube up which the nectar is sucked. Flowers that depend on pollination in order to reproduce are often a bright red in colour to attract hummingbirds, which are important agents of fertilisation, transferring pollen between flowers as they feed.

require a 'landing platform' on the edge of the flower. They hovered beneath a hanging blossom and probed for its store of nectar within while hanging in the air, thus appearing to defy gravity.

A female rufous-tailed hummingbird, not as active as the others in the area, perched on a twig beside her nest. The small home, built solely by her, was exquisitely formed of fine grasses, small leaves and moss, being lined with plant-down and cobwebs. She had completed the work a couple of days earlier and had spend much of her time smoothing the edges of the funnel-shaped nest as well as covering the outside with lichens and moss to give a neat finish.

Her feathers were ruffled and she was breathing quickly. A few minutes later she flew to her nest to stand in it restlessly, looking rather unwell. After three minutes she began to bear down, her feathers ruffling more as she did so. Within two minutes her bearing-down was occurring at five-second intervals and she pressed her vent occasionally against the sides of her nest, her tail feathers fanning out and quivering. Ten minutes later, she was labouring once a second, making a squeaking sound as she tried to lay her egg.

A sword-billed hummingbird flew about her nest to watch what was happening. Three minutes later, the egg was just visible but slipped back after each push the hen bird made. In another three minutes, the rufous-tailed hummingbird had worked her way to the edge of her nest until her head and the whole of the front of her body hung over its edge. She laboured a further two minutes, and the egg slipped clear of her vent to land neatly in the soft centre of the nest.

This rufous-breasted hummingbird is feeding her young by regurgitating food into their gaping mouths. Before mating she builds a delicate nest of plant material, fibres, moss, lichens and down and binds it together with silvery spider's webs. Despite its fragile appearance, it is both waterproof and solidly constructed even though it may be wound around hanging plants. Males of some species indulge in spectacular aerial displays to attract a female's attention before mating. Two white eggs are usually laid and are incubated by the female for two weeks before the chicks hatch. Although they are born naked they are fledged fully within three to four weeks and ready to leave the safety of the nest. Hummingbirds are fiercely territorial when nesting and use their superior speed and agility to chase away much larger birds which may venture close to the nest and its young.

As this hummingbird hovers by a flower its wings are beating between 20 and 70 times per second, too fast to be perceived by the naked eye although a high-speed camera can record the movements. The smaller the hummingbird, the faster its wings move in order to keep it airborne as it hovers, sometimes even upside-down. Throughout each narrow ellipse the wings are kept almost rigid and extended as they rotate at high speeds. The legs and feet of the hummingbird are not well-developed or strong and are used solely for perching, making the bird mainly aerial in its habits and lifestyle. Thus courtship and feeding both take place on the wing, and even small insects are caught and consumed in flight.

For ten minutes she rested, the egg warm beneath her, her eyes closed, before flying to a bromeliad plant where she drank some of the water cupped in the deep centre of its leaves. Other hummingbirds were hovering near her nest when she returned, showing their curiosity about the nest and its contents, but her darting display of annoyance drove them away to a safe distance. She resumed her watch on the egg. First she touched it gently with her bill, then nestled down on it. In three days' time she would lay a second egg to complete her clutch. Sixteen days later, when the naked, helpless black-skinned nestlings had broken free of the eggs, she would spend her days feeding them. She would poke her long, pointed beak delicately into their gaping mouths, each with a tiny stub of a bill at this stage, to inject the regurgitated food she had gathered for them. For 14 to 29 days she would work for her offspring with no help from the male. When their feathers had grown and they could leave the nest she might continue for a day or two to feed them on the wing, before they flew off to make their own adult lives away from her care and protection.

Her even more colourful mate, whose attentions to her had been confined to his courting and coupling with her to fertilize the egg, cavorted about the nest. His wings were blurred with the speed of their movement, leaving only his brilliant body visible like a tiny helicopter. He shot forward at about 80km per hour for 30m, made a right-angle turn with almost no change in speed, then decelerated to hover over the dripping leaves of a creeper. Landing lightly on the base of a curving leaf that glistened with water, he slid down to its tip, fluttering drops of water through his feathers with evident pleasure. He repeated the bath and, as he slid down another leaf, his colour changed dramatically. The lamellae on his feathers swelled invisibly to alter the wavelength of the light reflected from his iridescent plumage, turning the fire of his feathers greenish for a few moments. He slipped off the tip of the leaf and skidded helter-skelter down another leaf, fluttering all his feathers. He paused to wash his beak and rinse his tongue before whirring away to find a flower.

Fringing a thicket were some fine plants, hanging their red flower heads downwards. Hovering above one for a moment, the rufous-tailed hummingbird examined the flower head at the point where it joined the stem. He suddenly tilted, bill down. Still hovering, he began to spin, using his sharp bill to drill a neat hole through the calyx of the flower. Poised upside down and with his wings humming furiously, he stopped spinning to lap up the flower's nectar, swallowing the small insects that were also feeding on it.

A little way from him, a velvet-purple coronet hummingbird darted towards a fruit fly. The bird flew a little behind and below his prey before surging upwards to scoop it into his bill, but at the last moment the fly turned to one side and was struck only a glancing blow. It fell towards the ground, buzzing. But before it landed, the hummingbird made a twisting dive and caught it, less than a metre from the safety of the undergrowth.

The hummingbirds glided about the area of thin vegetation where the light, penetrating the thick jungle canopy, had ripened the blossoms. The birds seemed to move slowly, as if they swam in water, except when they changed direction. This they accomplished with such directness that it seemed as though an external and invisible force had flipped them from their chosen paths. Their wings were invisible in the dusky light as they hovered and skimmed from flower to flower.

Some 10m from the ground, a patrolling Julie's hummingbird (*Damophilia julie*) turned to rest on a budding twig. He shot forward towards the perch but before he reached it another hummingbird landed there. The Julie's hummingbird's wings hummed like an aeroplane engine as he hovered a little above the interloper. With a squeaking chirrup, he pushed forward his needle-sharp claws and punched them towards the perching bird who fluttered into the air to meet the challenge. The former bird's wings whirred even faster than before and he dashed himself against his foe to drive him backwards. The victim rolled on to his back, completed a loop and flew off, leaving the victor to occupy the twig. The skirmish had been briefer than is common among the aggressive species of hummingbirds, which often fight with the tenacity of terriers.

As the shafts of sunlight that pierced the jungle canopy grew weaker in the dying day, the hummingbirds flitting through the wet foliage became fewer. The hen birds that were incubating their eggs hovered beside their nests, carefully fitting themselves into the small openings and settling into the comfort of the downy interiors. The males and the females that had no nests or eggs to care for sank slowly with whirring wings on to twigs or branches and prepared to roost for the night. They ruffled out their feathers and, as they gave themselves up to deeper and deeper sleep, their heads tilted farther back and their tiny beaks pointed up towards the sky. With barely a heartbeat and imperceptible breath, they waited for the next day's sun to bring fresh fire to their feathers.

The ruby-topaz hummingbird of Brazil has a glossy body and wing plumage (left and below). Even unfamiliar flowers are thoroughly investigated by the hummingbird, whether they contain nectar or not. Hovering in front of the blossom, the tiny bird dives in and out of the flower to explore this new, exciting territory. The energy requirements of a hummingbird are enormous in proportion to its size, and it will visit up to 2000 flowers in a day to feed on their nectar and also to eat the insects that may be feeding in the flowers. The nectar provides the bird with essential carbohydrates and the insect food yields a source of protein. If a man functioned at the same rate as that of a hummingbird, he would require about 155,000 calories per day compared to an athlete's intake of 3000 calories. To supply this enormous energy demand the hummingbird must eat large amounts of food, even rich roods like nectar, which is made up principally of fructose and glucose, both extremely efficient energy foods. The bird's daily intake is equivalent to a quarter of its total body weight.

Rufous hummingbird

Distribution of hummingbirds

Hummingbirds are found only in the New World. Here, they are most abundant in the lush equatorial jungles, but they are found almost anywhere there are the flowers that provide the nectar for their diet. Ornithologists have spotted them seeking out blooms in the Andean forests as high as 5 000m up the snow line. The further one travels from the equator the fewer species one finds, but exceptionally hardy types live on the slopes of the mountain ranges of the western United States as far north as Alaska.

Most hummingbirds live in one territory all year round, but the northernmost species migrate southwards for the winter. The ruby-throated hummingbird, which occurs widely in the south-eastern parts of the United States, flies across the Gulf of Mexico to find warm weather and blossoms. Farther north, the rufous hummingbird leaves its home on the slopes of Mount Logan when the chill winds of autumn warn of the approach of winter, and begins its migration south.

These birds cross the coastal range of mountains, nearly 5 000m high, across their path. They fly over the states of Washington, Oregon and California, and navigate 800km of the Gulf of California to their winter home in Mexico. These incredible little birds cover the 3 200km of their journey despite mountains which force them to move up into cold turbulent air, and despite storms that fo them to fly down to sea-level, skimm through the wave troughs to avoid being ble far off course.

Before its migratory flight the humming gorges itself with food to increase its b bulk by more than half, living off this sto fat during the flight. In early spring the ru throated hummingbird puts on weight fo flight back to the fringes of the Arctic Cir The bird can put on this comparatively h amount of weight only twice a year, preparation for its migration. No one yet kn how this happens, but it has been sugge that a hormone signal produces the condi in the bird's body that allows it to accumu this energy fuel for the great effort to com

---

Hummingbirds are native to the New World a make their home wherever there is a plentiful supply of flowers which are rich in nectar. However, most species live in the tropical jung and rainforests near the equator, and there is steep fall in the varieties and their density in latitudes from 30° north and south of the equa The numbers on the map indicate the number species that have been observed in different ba of latitudes, and the arrows follow the directi of the routes of migrating species.

Opposite page:
Found in the forests of North and South America, there are 319 species of tiny hummingbirds which glow with colourful, iridescent plumage. In spite of their size, they are very aggressive, frequently attacking each other and even larger birds that trespass on to their territory. Some species migrate over great distances, the ruby-throated hummingbird flying non-stop across the 960km of the Gulf of Mexico in its southwards winter migration. The beaks vary from the slender bill of the ruby-topaz to the bizarre, elongated bill of the swordbill which is specially adapted for probing long, tubular flowers. The tails of different species also show variations and come in many shapes and sizes ranging from the streamer-type, long feathers of the ribbon-tailed hummingbird to the short, broad feathers of the magnificent rivolis. The Lodiges racket-tail has two outer, flowing tail feathers which end in large, rounded vanes.

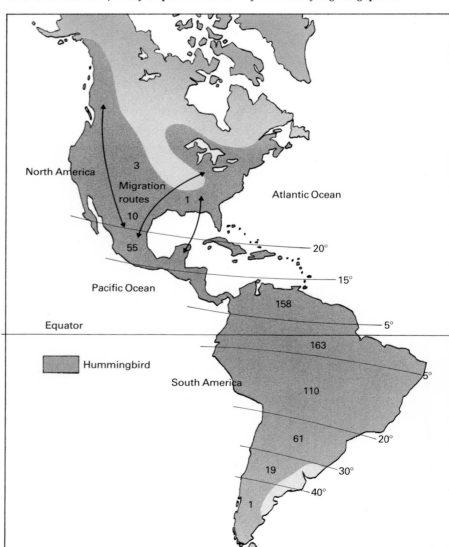

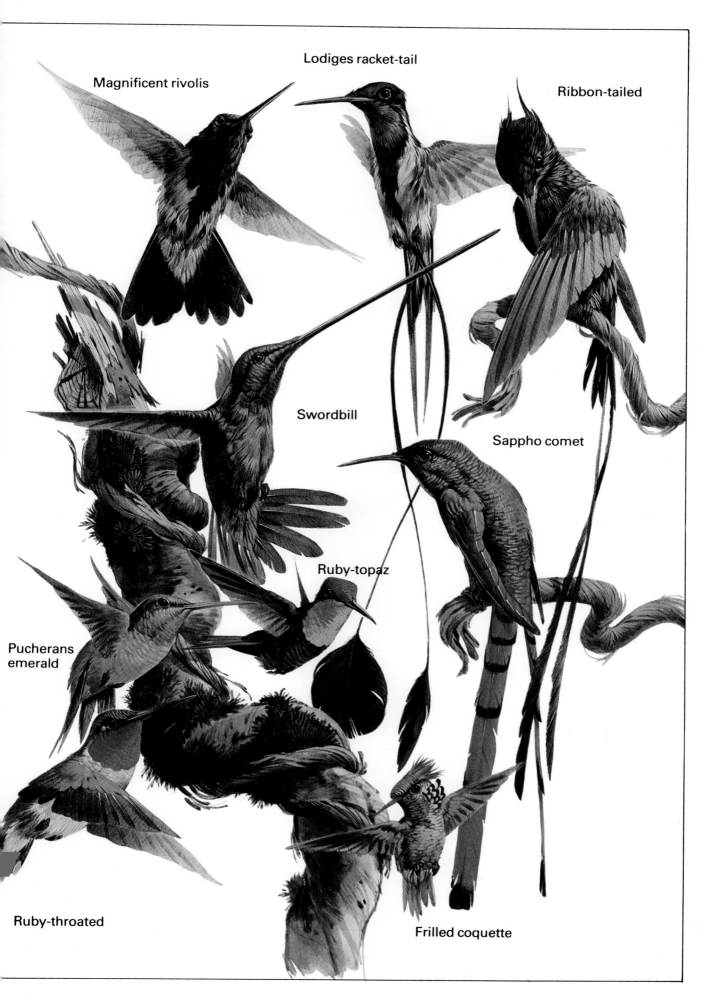

The hummingbird's flight

Hummingbirds are among the most accomplished fliers of the bird world; their capacity for aerobatics is greater than that of any other kind of bird. Not only can they fly fast, nearly 100km/h in level flight, and produce extremely rapid acceleration, but they also hover with a steadiness and precision that is beyond the attainments of any of the hovering hawks. Hummingbirds are among the very few birds indeed that can fly backwards and upside-down. Most birds dive by folding their wings and letting gravity do the work for them; the hummingbird is more ambitious and assists its fall with rapid wingbeats to exceed speeds of 100km/h. The secret of the hummingbird's flying superiority lies in the unusual structure of its wings and in its enormous output of energy. The ten preliminary feathers of the hummingbird's wing are strong and equipped with enlarged barbules which help to make the structure of the wing extremely rigid. The hummingbird's wingbeats are the most rapid of the bird world. The smaller the species of hummingbird, the faster the wingbeat. The small species will make as many as 80 beats per second, and it is this great speed that produces the characteristic humming sound that gives the bird its name.

The rainbow colours

The subdued plumage of the hummingbird leaps to life at the touch of light and movement. Their feathers are mostly pigmented with black and brown, but the barbules of the feathers are sheathed in microscopic scales of a horny material called keratin. Each scale, only 1/4000th of a millimetre in thickness, is clear and colourless, acting as a reflector in a similar way to the film of oil on the surface of a bubble, breaking light into its constituent prismatic colours. The lamellae (scales) are made up of many elliptical cells filled with air, and act like myriads of bubbles, breaking up the reflected light to make the bird a blaze of changing colour. The predominant colour of the iridescence is produced by the thickness of the lamellae reflecting light within a limited wavelength. The pigmented colours of the birds are produced by chemicals in the feather but these colours — though strong and clear — do not glisten with the iridescence of those that are scaled over with the minute lamellae.

At a glance	
Hummingbirds	
Phylum	Chordata
Class	Aves
Order	Apodiformes
Family	Trochilidae
Species	320 species of hummingbirds.
Smallest hummingbird	Princess Helen's hummingbird *(Calypte helenae)*
Body	50.8mm long
Bill	10mm long
Eggs	7.5mm long
Food	Nectar and insects.
Largest hummingbird	Giant hummingbird *(Patagonia gigas)*
Body	20cm long
Bill	4.25cm long
Food	Nectar and insects.
Eggs	2 in a clutch.

Hovering flight

Backwards flight

This sequence shows the wing movements of a hummingbird in flight and hovering in mid-air. In reality, the tiny bird's wings move so rapidly that it is impossible to observe the movement with the naked human eye and thus they may appear stationary apart from the incessant hum they make as they move. However, if the bird was seen in slow-motion these positions would be visible. Their manoeuvres in flight excel those of any other bird.

Sequence showing backwards somersault in flight
(Courtesy of American Museum of Natural History)

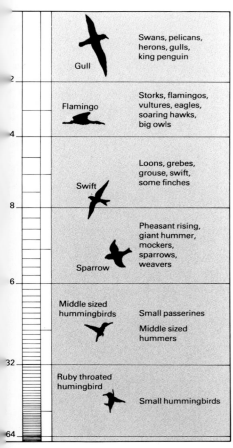

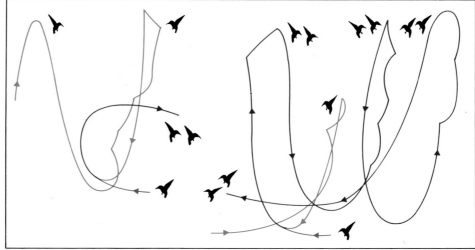

Many male hummingbirds indulge in unusual courtship displays to attract the attention of the females, but the aerial display of Anna's hummingbird is among the most spectacular. The male sings as he performs an elaborate series of arcs and loops consisting of symmetrical dives and ascents.

This chart compares the number of wingbeats per second of the hummingbird with those of other birds. Whereas the big pelicans and swans are leisurely fliers at one wing-beat per second, hummers may have 80 wingbeats.

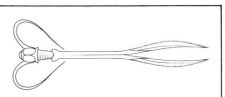

The hummingbird's tongue is uniquely adapted for extracting the nectar deep within flowers. An elongated, slender organ, it is divided into two narrow tubes which separate and curl inwards to form a funnel for extracting nectar.

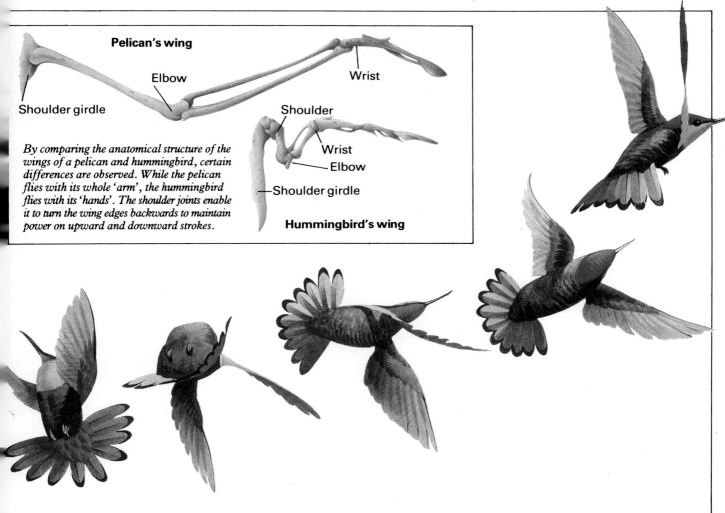

By comparing the anatomical structure of the wings of a pelican and hummingbird, certain differences are observed. While the pelican flies with its whole 'arm', the hummingbird flies with its 'hands'. The shoulder joints enable it to turn the wing edges backwards to maintain power on upward and downward strokes.

Picture Credits

Heather Angel
pages 41,46,47,93
Aquila
page 63
Ardea
pages 17,31,33,34,79,83,118,121
Bay Picture Library
pages 10,19,20,66,69,70,71,72,73
Bruce Coleman
endpapers and pages 30,32,35,94,95,100-101,121
Colour Library International
introductory page
Pat Morris
pages 16,62
National History Photographic Agency
pages 28,52,53,54,57,102-103,105,109
Nature Photographers
pages 80,81,82,84,85
Naturfoto
pages 110,111
Picturepoint
half-title page and contents page and pages 15,51
Satour
pages 56,59,78,104,107
Survival Anglia
page 107
Zefa
cover and pages 12,45,60,65,67,68,97,116,119